T0189994

Studies in Computational Intelligence

Volume 677

Series editor

Janusz Kacprzyk, Polish Academy of Sciences, Warsaw, Poland
e-mail: kacprzyk@ibspan.waw.pl

About this Series

The series "Studies in Computational Intelligence" (SCI) publishes new developments and advances in the various areas of computational intelligence—quickly and with a high quality. The intent is to cover the theory, applications, and design methods of computational intelligence, as embedded in the fields of engineering, computer science, physics and life sciences, as well as the methodologies behind them. The series contains monographs, lecture notes and edited volumes in computational intelligence spanning the areas of neural networks, connectionist systems, genetic algorithms, evolutionary computation, artificial intelligence, cellular automata, self-organizing systems, soft computing, fuzzy systems, and hybrid intelligent systems. Of particular value to both the contributors and the readership are the short publication timeframe and the worldwide distribution, which enable both wide and rapid dissemination of research output.

More information about this series at http://www.springer.com/series/7092

Leon R.A. Derczynski

Automatically Ordering Events and Times in Text

Leon R.A. Derczynski
Department of Computer Science
The University of Sheffield
Sheffield
UK

ISSN 1860-949X ISSN 1860-9503 (electronic)
Studies in Computational Intelligence
ISBN 978-3-319-83688-1 ISBN 978-3-319-47241-6 (eBook)
DOI 10.1007/978-3-319-47241-6

Softcover reprint of the hardcover 1st edition 2016

Printed on acid-free paper

This Springer imprint is published by Springer Nature
The registered company is Springer International Publishing AG
The registered company address is: Gewerbestrasse 11, 6330 Cham, Switzerland

To Nanna Inie; heartfelt thanks for your attention, focus and passion.

Foreword

I am delighted to be able to write a few words of introduction to this new book on time and language. It is published at a very important time, in the midst of an explosion in artificial intelligence, where humans, hardware, data, and methods have combined at a fantastic rate to help not only us, but also our tools and computers, better understand our world.

Across the globe, in almost every language we encounter, we discover that we have evolved the ability to reason about time. Terms such as 'now' and 'tomorrow' describe regions of time; other terms reference events, such as 'opened' or 'hurricane'. This ability to refer to times or to events through language is important and gives humans much great ability in planning, storytelling, and describing the world around us. However, referring to events and times is not quite enough—we also need to be able to describe how these pieces all fit together, so that we can say when an event, like the 'hurricane', happened. This temporal structure can be thought of being built from relations that link each event and each time like a net. These temporal relations are encoded in the way we use language around events and times. Discovering how that code works, and what temporal relations a text is communicating to us, is the key to understanding temporal structure in texts.

Traditionally, computational linguistics—the study of computational techniques for language—has given the tools used to address automatic extraction of temporal information from language. Temporal information extraction typically involves identifying events, identifying times, and trying to link them all together, following patterns and relations in the text. One of the harder parts of this extraction process is linking together of events and times, to understand temporal structure. There have been many clever approaches to the task, from scholars and researchers in industry around the world. It is so hard that there has been, and still is, a long-running set of shared exercises, just for this: the TempEval challenges. The first of this series was proposed almost a decade ago in 2006 by me and my collaborators, which we started in order to advance temporal semantic annotation and the plethora of surrounding tasks.

Later, it was actually through one of these TempEval tasks that I first met Dr. Derczynski, and thereafter over many coffees and late dinners at venues like

LREC, or ISA, the semantic annotation workshop. Since, we have collaborated on temporal information extraction, co-organizing more recent TempEval tasks. Our current forthcoming work is a full-length textbook with Marc Verhagen on temporal information processing, with plenty of examples and thorough discussion of the multitude of issues in this fascinating and open area of science.

However, despite our and the community's years of work, and the heavy focus of many researchers through shared task series such as TempEval and i2b2, the problem of extracting temporal structure remains one of the hardest to solve in extracting temporal structure, and also the most important. Clearly, some fresh knowledge is needed.

This book adopts a different tactic to many others' research and describes a data-driven approach to addressing the temporal structure extraction problem. Based on a temporal relation extraction exercise involving systems submitted by researchers across the world, the easy and difficult parts of temporal structure are separated. To tell us where the hardest parts of the problem are, there is an analysis of the temporal relations that few or even none of the systems get right. Part of this analysis then attributes to various sources of linguistic information regarding temporal structure. Each source of information is drawn from a different part of linguistics or philosophy, incorporating ideas of, for example, Vendler, Reichenbach, Allen, and Comrie. The analysis then drives into the later parts of the book, where different sources of temporal structure information are examined in turn. Each chapter discussing a source of this information goes on to present methods for using it in automatic extraction, and bringing it to bear on the core problem: getting the structure of times and events in text.

My hope with this line of work is that it will bring some new knowledge about what is really going on with how temporal relations related to language. We can see the many types of qualitative linguistic theoretical knowledge compared with the hard reality of computational systems' outputs of temporal relations, and firm links emerge between the two. For example, we see links between iconicity—the textual order of elements in a document—and temporal ordering; or, an elegant validation of Reichenbach's philosophically based tense calculus, which, by including the progressive, ends up at Freksa's formal semi-interval logic almost by accident, while continuing to be supported by corpus evidence.

Bringing together all these threads of knowledge about time in language, while coupling them with empirically supported methods and evidence from the data that we have, has been a fruitful activity. This book advances work on some big outstanding problems, raising many interesting research questions along the way for both computer science and linguistics. Most importantly, it represents a valuable contribution to temporal information extraction, and thus to our overall goal: understanding how to process our human language.

June 2016

James Pustejovsky
TJX/Feldberg Chair of Computer Science
Department of Computer Science
Volen Center for Complex Systems
Brandeis University
Arlington, MA

Acknowledgements

A very special thanks to Robert Gaizauskas for his extensive help and guidance at many points; and to Yorick Wilks and Mark Steedman for their comments on an earlier version. The whole could not have been possible without the vision for the field and vast groundwork laid describing time in language, which is due to in great part to James Pustejovsky, as is gratitude for the foreword.

Finally, the book was produced during a period where I received support from, in no particular order: the EC FP7 project TrendMiner, the EC FP7 project Pheme, the EC H2020 project Comrades, an EPSRC Enhanced Doctoral Training Grant, the University of Sheffield Engineering Researcher Society, and the CHIST-ERA EPSRC project uComp.

Contents

List of Figures

List of Tables

Abstract

The ability to describe the order of events is crucial for effective communication. It is used to describe causality, to plan, and to relay stories. This temporal ordering can be expressed linguistically in a variety of ways. For example, one may use tense to describe the relation between the time of speaking and other events, or use a temporal conjunction to temporally situate an event relative to time. This ordering remains the hardest task in processing time in text. Very sophisticated approaches have yielded only small improvements over initial attempts. This book covers relevant background and discusses the problem, and goes on to conduct an analysis of temporal ordering information. This breaks the types of information used into different groups. Two major sources of information are identified that provide typing information for two segments: relations explicitly described by a signal word, and relations involving a shift of tense and aspect. Following this, the book investigates automatic temporal relation typing in both these segments, presenting results, introducing new methods, and generating a set of new language resources.

Chapter 1
Introduction

Le temps mûrit toute choses; par le temps toutes choses viennent en évidence; le temps est père de la vérité.
Time ripens all things; with Time all things are revealed; Time is the father of truth.

Gargantua and Pantagruel
FRANCOIS RABELAIS

1.1 Setting the Scene

Humans developed natural language to communicate; over past millennia, it has been the most efficient form of transferring the majority of information between individuals. With the advent of computing, large amounts of natural language text are stored in digital format. The study of *computational linguistics* helps link the significant power of the computer with the efficiency of communicating in natural language.

Within computational linguistics, this research into identifying temporal information fits in the domain of *information extraction*. This sub-field concentrates on the automatic identification of specific information about entities, relations or events from natural language discourse [1]. It has developed to a point where information such as person-relation-data triples (such as *Carl Gustaf Folke Hubertus : job : King of Sweden*) can often be reliably identified [2]. The book concentrates on extracting information about temporal relations, which, as we shall see, is a challenging and difficult problem.

There is a strong need to understand time in discourse. Time is critical to our ability to communicate plans, stories and change. Further, much of the information and many of the assertions made in a text are bounded in time. For example, the sky was not always blue; George W. Bush's presidency was confined to an eight-year interval. An understanding of time in natural language text is critical to effective communication and must be accounted for in automatic processing and understanding of discourse.

Being able to identify times and events, the basic entities of temporal reasoning, within natural language discourse is not enough to understand its temporal structure. Events cannot independently be placed onto a calendar scale. To situate events

L.R.A. Derczynski, *Automatically Ordering Events and Times in Text*,
Studies in Computational Intelligence 677, DOI 10.1007/978-3-319-47241-6_1

temporally, they must be related to other events or to times. These relations, however interpreted, are the information describing the temporal structure of a text (corresponding to the C-series of McTaggart [3]: the fixed ordering of events); they allow one to situate an event in terms of times or other events and to describe the complexity of some event structures (for example, events with sub-events [4]).

Determining temporal relations is critical to understanding the temporal situation of events described in discourse. Automatic extraction of temporal relations has proven difficult, though it is something that human readers can perform very readily. Human readers are likely to have access to information from a given discourse as they read it and from experience in the form of world knowledge. That we can identify the nature of temporal relations easily suggests that the information required for temporal relation extraction is contained either in discourse or in world knowledge.

The task can be broken into two parts, given any document: identifying which phrases correspond to temporal entities in the document, such as times and events; and identifying how these entities are related to each other. Example 1 contains a selection of words and phrases of temporal relevance.

Example 1 **Nov. 17, 2006**$_1$ China's first ever space textbook **declassified**$_2$ and published.

Qian Xuesen's manuscript entitled "A General Introduction to the Missile" hit the shelves in Beijing on **Friday**$_3$, 50 years **after**$_4$ Qian first used it to **teach**$_5$ 156 university students, China's first generation of space scientists.

There are two times, (1) and (3); the first (1) is acting as the document's creation time. There are also examples of events, (2) and (5). Finally, there is a temporal signal (4), which explicitly describes the *type* of temporal relation that holds between a time (3) and an event (5).

Separate bodies of work focus on extracting *events* from texts and also on identifying and interpreting textual references to *times*. These have reached high recognition accuracies in both event and time recognition (see Sects. 2.2.3.3 and 2.3.3.3). However, identifying the types of relations that hold between these entities is still a difficult problem. After many years' effort, performance levels currently reach about 70 % accuracy (see Sect. 3.5).

Rather than attempt to further existing work on annotating individual events or times, the topic of this book is the temporal relations that hold between them. The task of determining which event and time pairs ought to be linked is referred to as the **relation identification** task. The problem of automatically determining the type of temporal relations that exist between a given pair of events or times is known as the **temporal relation typing** task. The following pages give particular focus to the temporal relation typing task.

Temporal relation typing consists of describing the kind of temporal ordering or relation between a pair of temporal entities, which are in themselves events or times. That is, deciding which order events happen in, according to the text. In Example 1, there are two events: entity 2, *declassified*, and entity 5, *teach*. To type the temporal relation between these two, we have to decide in which order they occur, according to the text. For this case, we would say that the *declassified* event happens after the *teach* event.

Finding that relation type just from the text is the relation typing task, and the focus of this work.

Once we can automatically find and label the temporal relations between events and times in a discourse, powerful techniques become available for improving human-computer interaction and automatic processing of information stored in natural language. For example, systems can perform better in answering questions put to them in natural language; it becomes easier to create better summaries of discourse; work on forensic applications – such as building timelines of witness statements in the event of a crime or transportation disaster – can be automated; one may construct stories using potentially incomplete accounts from multiple sources; and temporal information extraction could be used in everyday communication, to organise events and create calendar appointments automatically from personal communication. Work to solve the overall temporal information extraction problem continues in many specific areas of temporal information extraction, such as time and event recognition.

Considering temporal information allows better-informed processing of natural language. Understanding of temporal relations helps build rich and accurate models of information from discourse. For example, [5] use manually-added temporal information to augment kernel-based discourse relation recognition. Further, temporal expressions and relations often help not only with segmenting texts but also with relating matching segments [6–8]. Good temporal information extraction also provides clear benefits to summarisation and automatic biography systems (e.g. [9]), which require temporal information in order to determine clusters of events and a correct story order. Models of time in language have also been of use in machine translation [10]. Without an understanding of temporal relations in text, question answering systems cannot tackle any "when" questions or distinguish history and conjecture from current world state. In fact, every assertion is bounded in time and these bounds must be recognised in order to reason about events and world knowledge over time.

An example of the applications of temporal information in the NLP task of question answering follows. A system may be asked *"When was the current president of the USA elected?"*. Typically in question answering scenarios, a set of texts is provided as a basis for determining the correct answer. In this case, such texts may mention many presidents and have creation dates spread over many years. While statistical approaches using document timestamps have some success in answering temporal queries, better language understanding can provide reliable and precise results once the temporal information in discourse can be accurately interpreted [11, 12]. In this case, there may be multiple challenges: the current time must be determined, the current president must be identified, and the election event related to the identified president. Alternatively, one may simply look for the most recent match for this election event. Either way, an understanding of time in text is required [13]. To answer this correctly, systems may require not only the ability to situate a question in time and relate events described in a discourse to its creation date, but also distinguish mentions of historical events from more recent ones. Such a level of sophistication is not yet present in the state of the art of automatic temporal relation extraction.

Finally, prototype industrial applications have already been created that rely upon accurate relation of events and times in text. Carsim [14] extracts relations from car accident reports and uses the resulting information to construct 3D visualisations of potential stories leading to an accident. In the legal industry, temporal relation extraction is a critical part of verifying and merging independent accounts related to investigations [15]. Lastly, medical record processing involves a temporal aspect, allowing systems to automatically order events in a case history [16, 17].

The remainder of this chapter has three functions. The scope of the book is outlined, followed by a description of its key points. The chapter concludes with a brief outline of the book structure.

1.2 Aims and Objectives

How can we identify the information needed to determine the nature of a temporal relation and then use it to help automatically determine the nature of temporal relations? The previous section has described the kinds of temporal primitive used to convey temporal structure through natural language text, with a two-part general structure of (a) times and events and (b) relations between them. The automated processing of relations can be decomposed into two tasks. Firstly, the relation endpoints (individual textual references to times or events) are identified. Secondly, the nature of this binary relation is determined and described using a type from a pre-defined set of relations (containing concepts such as precedence, inclusion and identity).

For this book, we focus on the second task – determining the types of temporal relations – and do not attempt to determine which endpoints should be related (relation identification). Discovering which events or times might be related to each other is a difficult task to define; every time and event has temporal bounds, and so a temporal relation of some kind exists between all of them, though this may not be critical to the story told by any given discourse. Further, human-annotated datasets are available where the subjectively most salient binary temporal relations have already been identified. This allows us to focus on the pertinent and difficult task of determining the nature of relations, without worrying about the ill-defined task of how we should choose which binary relations to investigate.

The results is an evidence-backed, rational investigation into the difficulties of automatically understanding the temporal structure of a discourse. The key aims are:

1. To identify new information sources useful for temporal ordering;
2. To provide improved methods for extracting temporal ordering information;
3. To suggest avenues of further research in the area.

The work is constrained to English-language text in the newswire genre.

1.3 New Material in This Book

This book details an investigation into automatically determining temporal relations in text. This comprises a data-driven analysis of the temporal relation characterisation problem followed by two approaches to improving automatic performance at relation typing. The significant and novel contributions are described below.

The analysis draws upon new findings based on openly available datasets. It comprises the first analysis of temporal relation typing results from the TempEval-2 evaluation task [18], where many teams tested state-of-the-art temporal information extraction systems on a common set of data, including an attempt to automatically determine the nature of temporal relations. The analysis results in the definition of a "difficult" temporal relation, and identifies a consistently difficult set of temporal relations. These are the relations that must be conquered for research in the area to progress. Based on this set of difficult relations, there are quantitative and qualitative failure analyses. These explore a variety of linguistic phenomena related to temporality, and their prevalence among difficult temporal relations. Emerging from this new and detailed analysis, the section concludes with the presentation of evidence-based suggestions of directions for further research. Two of these are investigated in the following chapters.

The first approach for improving relation extraction is based on words and phrases that explicitly state the nature of a temporal relation – temporal signals. It begins with an empirical confirmation of signals as supporting temporal relations. Such confirmation is followed with the introduction and demonstration of a new technique, using signals to help in temporal relation typing, that achieves over 53 % error reduction when compared to the state of the art. As these signals are very useful, a corpus-driven characterisation of temporal signals is given, the first of its kind, including statistical results and a formal definition of this closed class. It is found that these signals exhibit two kinds of polysemy, and quantitative results are presented describing both of these. Firstly, a linguistic polysemy where the signal words and phrases have both temporal and non-temporal meanings. Secondly, a temporal polysemy, where the same signals may have differing temporal interpretations depending on their context. After this characterisation, existing resources are curated and augmented to create a temporally annotated corpus with high quality signal annotations, which is presented as a new linguistic resource. Having been shown to be useful but ambiguous, we introduce a two-stage process for annotating temporal signals. The initial part of this is an effective method for automatic identification of signals, which is similar to a specialised word sense disambiguation task. The remaining part of the process is an effective method for, given a word or phrase that acts as a temporal signal, associating signals with their event or time arguments, thus connecting a signal to a binary temporal relation. This final part entails the definition of a new discourse-based task, and proposes multiple approaches, arriving at a successful initial solution. The approach concludes with a demonstration that these new techniques for identification of signals and association with their arguments, coupled with our approach for using them to support relation extraction, improves on the overall characterisation of temporal relations within a given discourse.

The second approach for improving relation extraction centres upon a framework of tense and aspect than can deterministically provide basic temporal ordering information between some events and times. Because the framework is not directly applicable to existing resources, two interpretations for the framework are put forward. The subsequent investigation begins with the first corpus-driven validation of this established framework of tense and aspect. The validation empirically demonstrates that the model is consistent with a gold standard temporally annotated corpus, but that finding which events and times are connected is an open problem. To confirm the results of this validation, the model's predictions are for the first time integrated into machine learning approaches for describing temporal relations. This gives improvements for event-event relation typing. It is also shown that the model has utility for determining the nature of relations between times and events. A technique is presented for using the model to automatically classify the temporal relations between times and events. The problem of determining which events and times to consider for connection under this framework is shown to be a limiting factor in its application. Finally, an ISO-compliant mark-up for integrating this model with established temporal annotation schemas is presented to aid further work using the model, parts of which have been shown to be helpful for event-time relation extraction and to be critical to the understanding of some other temporal phenomena.

1.4 Structure of the Book

This book details a plan for improving temporal relation typing, and describes the outcome of the plan. It finds that to understand the temporal ordering of events described in text we cannot rely on a single set of information for every relation. Instead, we need to draw upon multiple heterogeneous information sources. A set of these information sources is identified and techniques introduced for exploiting them to improve temporal relation typing. Two of these information sources – one related to explicit temporal signal words and phrases, another related to tense and aspect – are explored in depth.

The remainder of this document is divided into three major sections. Early chapters introduce the field and prior work. The next set of chapters comprise the core of the work and describe the experimental approach and its results. Finally, an overview is given, discussing applications of temporal relation typing and providing conclusions. Appendices contain supplementary material.

Chapter 2 describes necessary theoretical and computational background. Chapter 3 explores related work on the specific task of temporal relation extraction, concluding with the current state of the art. It also briefly introduces current event and time extraction systems.

The temporal relation extraction problem is detailed in depth in Chap. 4, which also includes failure analysis of a set of temporal relation classifiers and outlines our approach for the remainder of the experimental work. Chapter 5 introduces temporal signals, showing how they are useful for relation typing and how they can be automatically annotated in a helpful manner. Chapter 6 details a model of tense which is used to improve relation typing between verb events.

Finally, Chap. 7 provides a formal summary of the book and a discussion of promising future directions for temporal relation typing and the overall task of temporal information extraction.

References

1. Gaizauskas, R., Wilks, Y.: Information extraction: beyond document retrieval. J. Doc. **54**(1), 70–105 (1998)
2. Carlson, A., Betteridge, J., Wang, R., Hruschka Jr, E., Mitchell, T.: Coupled semi-supervised learning for information extraction. In: Proceedings of the third ACM International Conference on Web search and data mining, pp. 101–110 (2010)
3. McTaggart, J.: The unreality of time. Mind **17**(4), 457 (1908)
4. Pustejovsky, J.: The syntax of event structure. Cognition **41**(1–3), 47 (1991)
5. Wang, W., Su, J., Tan, C.: Kernel based discourse relation recognition with temporal ordering information. In: Proceedings of the Annual Meeting of the Association for Computational Linguistics, pp. 710–719 (2010)
6. Bestgen, Y., Vonk, W.: Temporal adverbials as segmentation markers in discourse comprehension. J. Mem. Lang. **42**(1), 74–87 (1999)
7. Bramsen, P., Deshpande, P., Lee, Y., Barzilay, R.: Finding temporal order in discharge summaries. In: AMIA Annual Symposium Proceedings, American Medical Informatics Association vol. 2006, p. 81 (2006)
8. Jean-Louis, L., Besançon, R., Ferret, O.: Using temporal cues for segmenting texts into events. Adv. Nat. Lang. Process. **6223**, 150–161 (2010)
9. Filatova, E., Hatzivassiloglou, V.: Event-based extractive summarization. In: Proceedings of the ACL Workshop on Summarization, pp. 104–111 (2004)
10. Horie, A., Tanaka-Ishii, K., Ishizuka, M.: Verb temporality analysis using Reichenbach's tense system: Towards interlingual MT. In: Proceedings of the International Conference on Computational Linguistics, Association for Computational Linguistics pp. 471–482 (2012)
11. Derczynski, L.R., Yang, B., Jensen, C.S.: Towards context-aware search and analysis on social media data. In: Proceedings of the 16th International Conference on extending database technology, ACM, pp. 137–142 (2013)
12. Derczynski, L., Gaizauskas, R.: Information retrieval for temporal bounding. In: Proceedings of the 2013 Conference on the Theory of Information Retrieval, ACM, pp. 129–130 (2013)
13. Derczynski, L., Strötgen, J., Campos, R., Alonso, O.: Time and information retrieval: introduction to the special issue. Inf. Process. Manag. **51**(6), 786–790 (2015)
14. Johansson, R., Berglund, A., Danielsson, M., Nugues, P.: Automatic text-to-scene conversion in the traffic accident domain. In: International Joint Conference on Artificial Intelligence, vol. 19, p. 1073 (2005)
15. Howald, B., Katz, E.: On the explicit and implicit spatiotemporal architecture of narratives of personal experience. Spat. Inf. Theory **6899**, 434–454 (2011)
16. Savova, G., Bethard, S., Styler, W., Martin, J., Palmer, M., Masanz, J., Ward, W.: Towards temporal relation discovery from the clinical narrative. In: AMIA Annual Symposium Proceedings, American Medical Informatics Association, vol. 2009, p. 568 (2009)
17. Jung, H., Allen, J., Blaylock, N., de Beaumont, W., Galescu, L., Swift, M.: Building timelines from narrative clinical records: Initial results based-on deep natural language understanding. In: Proceedings of ACL-HLT, Association for Computational Linguistics, p. 146 (2011)

18. Verhagen, M., Saurí, R., Caselli, T., Pustejovsky, J.: SemEval-2010 task 13: TempEval-2. In: Proceedings of the 5th International Workshop on Semantic Evaluation, Association for Computational Linguistics, pp. 57–62 (2010)
19. Derczynski, L.: Determining the types of temporal relations in discourse. Ph.D. thesis, The University of Sheffield (2013)

Chapter 2
Events and Times

Hvor er jeg? Hvad vil det sige:Verden? Hvad betyder dette Ord?
Hvo har narret mig ind i det Hele, og lader mig nu staae der?
Where am I? What does it mean to say: the world? What is the meaning of that word? Who tricked me into this whole thing and leaves me standing here?

Repetition
SØREN KIERKEGAARD

2.1 Introduction

Time is a critical part of language. Without the ability to express it, we cannot plan, tell stories or discuss change. Almost all empirical assertions are transient and have temporal bounds; because of this the capability to describe the future, the past and the present is critical to accurate information transfer through language.

If we are to have a computer reason about times and events, we need to know about time in language. Time in language can be broken down into three primitives: times, events and temporal relations [1]. Viewing the temporal structure of a discourse as a graph, the times and events are the nodes and the relations the arcs. In this chapter, we introduce the nodes – events and times.

Some theories and models of language include or focus on temporality. While some linguistic theories related to time require a human-level understanding of text, others use very finite terms which operate using features of language that we can already automatically identify with a high degree of confidence. Based on these linguistic theories, we can describe certain structures in text as well as their behaviour. We may leverage this to better understand and process temporal information in discourse.

Finally, given this understanding, it is possible to build systems for some automatic temporal processing. There are approaches to detecting times and events, to determining event durations [2, 3] and to typing the relation between two events [4–6].

L.R.A. Derczynski, *Automatically Ordering Events and Times in Text*,
Studies in Computational Intelligence 677, DOI 10.1007/978-3-319-47241-6_2

This chapter presents background material relevant to time in language. It first discusses events and then times, covering for both the issues of definition, annotation and automatic processing.

2.2 Events

The Oxford English Dictionary defines an event as *"a thing that happens or takes place, especially one of importance"*. This definition could be broken down into occasions, actions, occurrences and states. However, the occasions, actions, occurrences and states are used in natural language more widely than this definition permits; there are often mentions of negated events, conditional events or modal events, which cannot be said to certainly "happen or take place" [7]. Further, events can be composed of many sub-events: for example, the Arab Spring lasted months and included multiple revolutions, each of which had a long history, a complex set of story threads all happening in parallel, a culmination and an aftermath. Indeed, processing historical events has its own challenges [8]. In addition, the definition of an event mention varies. Events may be represented by a variety of lengths of expressions, ranging from document collections [9] to single tokens. For the purpose of this book, the description of events from TimeML (a temporal markup language, [10]) is adopted, as follows:

> We consider "events" a cover term for situations that happen or occur. Events can be punctual or last for a period of time. We also consider as events those predicates describing states or circumstances in which something obtains or holds true.

Given that they may describe an action or transition, events are often expressed by verbs (*"The bus stopped suddenly"*). A nominalised event is an event that is represented by a noun phrase. For example, one might mention *the explosion*, which is a noun that describes an event. Events may also be expressed by statives, as in *the man was an idiot*); by predicatives, as in *Elizabeth is queen*; by adjectives, in *the storm is active*; and by prepositional phrases, such as in *soldiers will be present in uniform*. A further discussion of events and states can be found in [11].

Events do not have to be real and observable for them to be annotated in a given text. Unreal events, such as those in a fictional or modal context should be included in a temporal annotation of a document. Description of future events or of things subordinated into the conditional world of an *if* (for example) are still events, and ought to be processed as such.

2.2.1 Types of Event

Independent of their form of expression, events may be taxonomised into discrete classes. These are introduced as follows.

Occurrences

These denote something factual that happens or occurs. The event is not modal or intensional, and the account of the event is given first-hand. For example, *There was an explosion shortly before 11a.m.*.

Reports

These events are those of some actor relaying information about other events or states. The actor may be declaring, narrating, commenting upon or otherwise reporting. Typically in English, this class of events is expressed with words such as *said*, *told* and *explained*.

Perceptions

In some contrast to reports, perceptions are events that describe the observation or capture of some other event. Typical words that might be used for events in this class include *hear*, *see* and *discover*.

States

This class of events introduces something that holds true, such as an observation about world state.

Intensional Actions

These involve some actor with a specific (perhaps unstated) goal in mind, who performs distinct actions following that intent. The event is the expression of intentionality. Examples include *Microsoft tried to monopolize internet access.*

Aspectual

Finally, aspectual events are those expressions which describe certain parts of the life of an event, such as its beginning, culmination, continuation and so on. For example, *The scientists were starting to show signs of exhaustion*. See also [12].

While it is possible to sometimes further sub-categorise events, or group them in other ways, this coarse separation of event classes is ample for the scope of this book.

2.2.2 Schema for Event Annotation

Given definitions of events and a need to process them automatically, some kind of formal method of describing events must be introduced. For this, and for temporal annotation over the remainder of this book, we adopt TimeML. TimeML [10] is an XML-style markup for temporal information in natural language texts and has become an ISO standard. An overview of the syntax and annotation guidelines can be found online.[1]

[1] See http://www.timeml.org/.

TimeML proposes annotating events expressed in text with the <EVENT> tag, which has an `class` attribute. The `class` attribute contains one of a set range of values, depending on the class the event belongs to. TimeML's event class taxonomy is slightly richer than the one described above but essentially similar.

It is important to determine exactly what to annotate. Events may have actors, for example, and may be expressed using auxiliaries, prepositional phrases, negation and modal signifiers, and so on. The contiguous sequence of words that describes an event is called the event chunk. The single most important word within this chunk – the one that critically defines the event, such as the dominant verb – is the chunk (or event) head. In TimeML, <EVENT> annotations are applied to the shortest possible phrase that could describe the event; e.g., its head. See Example 2 from the TimeML 1.2.1 annotation guidelines.

Example 2 He would not have been going to permit anything like that.

In the example, negation, modality and an auxiliary-based tense structure are applied to the event, but only the head of the phrase is to be annotated.

TimeML also allows the annotation of extra information regarding events. This information may not be critical to the temporal significance of the event, but is certainly of linguistic interest and has proven helpful to many automatic annotation systems. The auxiliary attributes available are rough guidelines, rather than a precise or exhaustive set of temporal facets of events. Attributes of events annotated include:

- Part of speech (noun, verb etc.);
- Tense, from a limited set of values;
- Aspect, covering progressiveness and perfectiveness;
- Cardinality, indicating how many times the event may have been repeated;
- Polarity, to capture negation;
- Modality, holding the type of modality (if any) that applies to the event.

2.2.3 Automatic Event Annotation

2.2.3.1 Task Description

Complete event annotation comprises **event recognition** (determining which expressions denote events) and **event classification** (characterising events once found). Recognition concerns determining which words or phrases can be marked up as being events. Event classification involves determining the "class" of a particular event (such as an action or a state) according to a schema such as that presented in Sect. 2.2.1. Performing both tasks together is generally harder than just recognising where events lie in text [13].

2.2.3.2 Evaluation

In automatic event annotation, both recognition and classification of events need to be evaluated. Firstly, it should be possible to score a system's performance at identifying the textual extents of event words or phrases. Secondly, the assigned class of an event needs to be evaluated. This can be done with a simple correct or incorrect choice, leading to an overall accuracy score for a set of event class assignments.

Identifying Event Extents

Event recognition is the task of identifying and delimiting event phrases. A perfect system will mark all events, determining their textual bounds correctly and not mark any text that is not an event. Evaluation metrics should thus reward systems for both finding events and also for not finding non-events. Precision and recall fit these requirements and are often used to evaluate event recognition [14]. A brief description of precision and recall follows.

Recall is the proportion of existing items that have been identified by a system; a system that returns one event in a document that actually contains ten has a recall of 10 %. However, a system that marks everything as an event is bound to find all events and has a recall of 100 %. To balance this, one may introduce precision. Precision is the proportion of returned items that are correct; returning just one correct item and no others gives 100 % precision, but returning everything where there are only a few events will generate a low precision score.

Assuming events are always exactly one word long, if W is the set of identified words and E is the set of words that are events, we can define precision and recall as follows.

$$recall\ R = \frac{W \cap E}{W} \tag{2.1}$$

$$precision\ P = \frac{W \cap E}{E} \tag{2.2}$$

Relations between precision and recall are discussed by [15]. It is common to combine the two with a harmonic mean such as F-measure [16]. The formula is as follows:

$$F_\beta = (1 + \beta^2)\frac{PR}{\beta^2 P + R} \tag{2.3}$$

This is also known as the F1 score. The "1" in F1 corresponds to a weighting between precision and recall, with them being equal. A flexible F_β measure is also available, with low β favouring precision and high β favouring recall. A β of 0.5 may be desirable if one wants to particularly penalise spurious event annotations.

2.2.3.3 Approaches

Recognising and annotating event mentions in text has been approached in a variety of ways. It has been approach in a variety of ways, cast separately as a named entity recognition problem or as a syntactic analysis problem. The current most successful approaches combine both these approaches, and use semantic role information to reach comparatively high performance.

Boguraev et al. [13] cast TimeML EVENT recognition as a machine learning chunking problem. Text is treated as a sequence of tokens to which labels are assigned which describe chunk boundary information; three labels are possible – E for an end of a chunk, I for a token inside a chunk and O for "any token outside a target chunk". Features are then generated based on capitalisation, n-gram, part of speech, chunk type and head word information, similar to a word-profiling approach to entity recognition [17]. Following this, recognising EVENT extents in the Wall Street Journal is 77–80% accurate (F-measure). This figure drops to 61–64% accuracy for the joint task of recognising event extents and then correctly assigning TimeML classes to these events. The difference shows that the event classification task is non-trivial, having similar success rates to the approach used here for event recognition (e.g. around 75–80%).

EVITA [18], included in the TARSQI toolkit (Section A.3.1), employs different strategies for dealing with verb, noun and adjective events. It uses both machine learning and knowledge-based techniques. Verbs are filtered based on the verbal chunk head, modal auxiliaries and event polarity. Nouns are filtered against a look-up table and sense disambiguation lookup (to repeat the example from the paper, a noun in WordNet synset *phenomenon* is not an event if is it also subsumed by the synset *cloud*). Finally, adjectives are only tagged as events if they have already been used as such by a gold standard source (such as TimeBank). EVITA reaches 80 % F-measure when recognising verbal events in TimeBank 1.2, which is comparable to IAA scores from that corpus' creation.

More recent efforts in automatic TimeML event annotation focus on machine learning approaches incorporating information about semantic roles, reaching F-measures of over 0.80. One leading tool, TIPSem-B [19], incorporates semantic role information into its CRF-based event annotation approach. It is openly available for download.[2] Other approaches have refreshed existing systems like EVITA and included the whole into common NLP frameworks; for example, GATE-Time [20] adapts EVITA into a machine learning system as a GATE component, making it easy to port between applications and capable of using extra training data to improve performance.

[2] See http://gplsi.dlsi.ua.es/demos/TIMEE/.

2.3 Temporal Expressions

Temporal information in text is often expressed using a phrase that precisely describes a point or duration. Sometimes these points reference an absolute unambiguous time (anchored via e.g. a calendar), which is of great help when trying to map events from a discourse to a timeline. It is also often that case that such phrases explicitly state an interval's length. Because they are so explicit, these phrases are used when temporality is critical. Thus, attempts to extract a discourse's temporal information must capture and process these phrases.

Linguistic characterisation of temporal expressions has led to discussion and observations regarding their usage and situation. Hitzeman [21] found that time expressions are often used as discourse segmentation markers and highlights their potential ambiguity. They find that the interpretation of a given temporal expression depends on its syntactic position. Similarly, Bestgen and Vonk [22] show that temporal expressions used as adverbials help set the scene for a sub-part of discourse, providing a context and a timeframe and are helpful discourse segmentation markers, improving discourse comprehension. Cohen and Schwer [23] perform multi-lingual characterisation of temporal markers, describing such expressions as comprising three parts: the size of the temporal segment, the distance from a temporal centre (e.g. a reference point, a concept addressed in detail during Sect. 6.3.1) and an orientation such as future or past. Finally, [24] is entirely dedicated to temporal expressions and the current reference book on the topic.

For this book, a "temporal expression", or **timex**, is any expression that denotes a moment, interval or other temporal region without having to rely upon an event. Each interval is composed of two points between which it obtains. For example, *24th August 1997*, *two weeks* and *now* are all temporal expressions; *after the storm* is not. Hobbs and Pan [25] define a "proper interval" as one where the start point is before the end point. Under this definition, this book considers only "proper interval" as intervals; that is, no minimum atomic duration is recognised, and there is no quantisation of time into chronons. Rather, temporal entities are described by infinitesimal points that bound them.

One needs to discover where these expressions occur in text and understand something of their semantics before being able to connect them using temporal relations.

2.3.1 *Temporal Expression Types*

Before describing algorithms that can identify and anchor time expressions, we will briefly equip the reader with a short summary of types of time expression. Most papers that cover this topic, using varying nomenclatures, settle on a small set of different types of time expressions defined by their authors [26–30]. These types can generally be mapped onto one of the following distinct classes.

- **Absolute** — Where the text explicitly states an unambiguous time. Depending on the granularity of the interval, the text includes enough information to narrow a point or interval directly down to one single occurrence. This is in contrast to a time which, while precise and maybe easy for humans to pin onto a calendar, relies on an external reference. For example, the interval *Thursday October 1st, 2009* would be considered absolute, but *The week after next* would not – the information is not all explicit or held in the same place; this latter expression implies reliance on some external reference time. Absolute expressions are sometimes also known as fully-qualified time expressions.
- **Deictic** — Cases where, given a known time of utterance, one can determine the period being referred to. These time expressions, specify a temporal distance and direction from the utterance time. One might see a magazine bulletin begin *Two weeks ago, we were still in Saigon.*; this expression leaves an unclear implicit speech time, which one could safely assume was the date the article was written. More common examples include *tomorrow* and *yesterday*, which are both offset from speech time; to describe this using Reichenbach's model (Sect. 6.3), deictic temporal expressions represent situations where reference time and speech time are the same.
- **Anaphoric** — Where speech and reference time are not at the same point. Anaphoric temporal expressions have three parts – temporal distance (e.g. 4 days), temporal direction (past or future) and an anchor that the distance and direction are applied from. The anchor, for anaphoric temporal expressions, is the current reference time as per Reichenbach's model (Sect. 6.3). Example phrases include *the next week*, *that evening* or *a few hours later*, none of which can be anchored even when their speech time is known.
- **Duration** — A duration describes an interval bounded by a start and an end, where the distance between the two is known, but the expression itself is not placeable on any external time system (like a calendar). Durations generally include a time unit as their head token; for example, *ninety minutes* is a single duration timex. This type of temporal expression is easily confused with deictic expressions; to use Ahn's example [28],

Example 3 "In the sentence *The Texas Seven hid out there for three weeks*, the timex *three weeks* refers to a duration, whereas in the sentence *California may run out of cash in three weeks*, the same timex refers to a point three weeks after the reference point".

- **Set** — Regularly recurring times, such as "every Christmas" or "each Tuesday". These usually have a regular interval between occurrences and persist for a duration or describe a point event ("every other Thursday at 4.30pm").

2.3.2 *Schema for Timex Annotation*

Temporal expressions are often inherently vague, and typically only communicated only to the level of precision that the speaker requires in order to convey their point coherently. As a result, it is difficult to develop a precise, discrete knowledge representation form for timexes – the classic AI problem of building machine-readable forms from qualitative concepts. Bearing this in mind, approaches to timex annotation have been developed.

Direct anchoring points for times and events comprise normalised temporal expressions – that is, linguistic expressions that refer to a time, which can be placed onto an absolute calendar scale. For example, *"2 July 2009"* is an unambiguous date. Some reasoning may be required in order to normalise a temporal expression; one may encounter text such as *"on Sunday"*, which requires a reference temporal expression that is better specified before it can be absolutely positioned. The recognition, categorisation and normalisation of temporal expressions is briefly discussed in Sect. 2.3.

To this end, any timex annotation schema has to account for describing both the extents of the expression and the value of the expression itself. Today, the two prevailing standards for timex annotation are TIMEX2 and TIMEX3. These standards evolved through the MUC [31] exercises and TERN [13] through TIMEX to more recent incarnations. Both are XML-based and cater for the timex classes of duration, time, date and set.

An annotation schema should provide a way of marking up events, times and relations in text. Additional information can be provided, such as normalisations of times, tense and aspect information, markup of temporal signals such time adverbials, aspectual links and so on. This book works with the TimeML annotation standard, as it is the most active and has the largest amount of annotated resources. TimeML accounts for not only timexes but also event and temporal relation annotation. Only the timex aspects are discussed in this section.

This section introduces the TimeML, TIMEX and TCNL annotation schemas. Other notations are available, but as the future work in this book concentrates on TimeML, an exhaustive cataloguing would not be appropriate.

2.3.2.1 TimeML

TimeML [10] is an XML-based language for temporal annotation. It allows annotation of events and times, with a rich format for each, as well as thorough provision of links to capture relations between events and times:

- TLINK: temporal, possibly including references to supporting words
- SLINK: subordinate, for modality, evidentials and factives
- ALINK: aspectual, only between two events, describing an aspectual connection

As well as this, TimeML includes a comprehensive event annotation and uses the TIMEX3 standard described above for representing temporal expressions. One may also link signals (such as temporal adverbials) with events or temporal links, to show sources of temporal information in text. TimeML is the only temporal annotation language to become an ISO standard.[3] Widespread adoption has lead to many temporal information extraction experiments using TimeML annotated corpora, as well as multiple iterations of the language and the production of processing tools that can parse the markup.

TimeML does not employ the Allen interval relations, but instead uses its own set, based on Allen's earlier work [32, 33]. Notably, TimeML has no OVERLAPS relation, or way of expressing it. This is clarified in TimeML-strict [34]. A fuller introduction to TimeML can be found in [35].

ISO-TimeML [36] is a LAF and TEI compatible iteration of TimeML. It permits stand-off annotation, where the SGML annotations do not clutter text by being inline and has a more elegant method of instantiating events. The formal standard is recognised by the ISO and maintained by an active working committee.

2.3.2.2 TIMEX3

TIMEX3 stipulates the annotation of smaller strings than TIMEX2 [27] and is intended for use alongside mechanisms for annotating temporal links and events. TIMEX2 permits longer expressions, including event-based timexes which are anchored not to absolute scales but to events described in the text; the rationale behind this is that TIMEX2 was not designed for use in an environment that included event annotations, whereas TIMEX3 is intended to be used as part of the TimeML annotation scheme. TIMEX3 is designed to work across domains [37]. Focused research further details the differences between the two standards and describes an approach for converting data from the TIMEX2 to TIMEX3 standard [30].

TIMEX3 is currently used as the means of describing times in TimeML; it looks like this:

```
<TIMEX3
        tid="t43" type="DATE"
        value="1989-10-30"
        functionInDocument="CREATION_TIME">
                10/30/89
</TIMEX3>
```

The `value` field may take the form an ISO8601-format date, a `P` followed by a numeric quantity and unit symbol to denote a period, or one of a number of special anaphoric-based values such as `PRESENT_REF`. Its format is not trivial and the TIDES/TimeML documentation are the best resources for its description [10, 27]. For the scope of this book, we generally consider TIMEX3 in the context of

[3] ISO WD 24617-1:2007

TimeML, as we are interested in an annotation schema that covers not only temporal expressions but also events and temporal relations. Other attributes of TIMEX3 annotations include:

- Function in document, to denote special timexes such as the document creation point;
- Type, to capture the timex class;
- Modifier for adding information that cannot be added to the value, such as qualitative information (e.g. *"the dawn of 2000"* would be marked as the year 2000 with a modifier of `start`);
- Quantifier and frequency for describing the repetition pattern of a set timex; for example, *"every other Sunday"* would have a value of `P1W` (period of 1 week) and a quantifier of `every other` – and *"twice a day"* has value `P1D` with a frequency of 2.

2.3.2.3 TCNL

TCNL [29] is "a compact representational language" for working with time expressions. A set of operators and labels are defined, which can be combined to produce various offsets or absolute expressions. For example, TCNL looks elegant for simplistic temporal relations; $\{tue, < \{|25\{day\}|@\{dec\}\}\}$ for *Tuesday before Christmas*, or $\{friday, < now\}$ to represent an earlier Friday. A calendar model, working with different levels of granularity, is used to help anchor times. Weeks and months, for example, have different durations and do not share synchronised boundaries, but both – when combined with an integer – can define a solidly bounded absolute interval; e.g. Week 34 2008, or January 2012.

Its authors suggest that TCNL has benefits over TOP [38], TIMEX2 and TIMEX3/TimeML; namely, that TCNL is calendar-agnostic, focuses on intensional meaning of expressions (which are allowed in TimeML, but not compulsory and not used in the two largest TimeML corpora), shows contextual dependency by using references such as focus and that its type system makes granularity conversion transparent.

An example of TCNL's capture of intensionsal time reference – "Yesterday" becomes $\{now - |1day|\}$ instead of something like 20090506. A set of operators are used to reason between operands:

- $+/-$ for forward/reverse shifting.
- @ for in; e.g., $\{|2sun|@\{may\}\}$ is "the second Sunday in May".
- & for distribution; e.g., $\{15hour\}\&[\{wed\} : \{fri\}]\}$ is "3pm from Wednesday to Friday".

Performing some basic algebra, "Friday last week" is split, into "Friday" and "last week". This is represented thus:

$\{fri\} + \{now - |1week|\} = \{fri, \{now - |1week|\}\} = \{now - |1fri|\}$

Further examples in TCNL and a reference guide to the language, can be found in [39].

2.3.3 *Automatic Timex Annotation*

2.3.3.1 Task Description

As with events, extracting timexes can be decomposed into a multi-part task. In this case, the principal parts are determining which words and phrases in a document comprise timexes and then assigning various attribute values to that phrase. Once identified, a temporal expression may be converted to a fully specified date or interval. Existing work has investigated the task of "anchoring" or "normalising" temporal expressions; that is, taking tokens from a document and mapping them to an absolute temporal scale, using an agreed notation. For example, while the single 24-hour period that the expression *next Thursday* refers may be immediately clear to us at any given time, some processing is required on the part of a computer to map this into a complete time (specifying at least a year and day). It is also important to choose the correct granularity for temporal expressions; *next day* refers loosely to the contents of a 24-hour period, not to a period of precisely 86400 seconds occurring between two local midnights (or however many caesium decay events, in SI terms).

Automatic timex annotation is typically a three-stage process. Firstly, one must determine the extents of a temporal expression. This stage may be evaluated using conventional precision and recall measures. Secondly, the timex should be interpreted [40], converting it to a representation according to an established convention. This includes assigning both an expression type and value, which can be evaluated with string matching for strict evaluation. Thirdly and optionally, the timex may be anchored to a time scale, which involves mapping it to a specific time, date, or range of times and dates.

Even in the case of temporal expressions, apart from those that are absolutely anchored in text – that is, those that include a year placed along an agreed calendar system – one will have to use some knowledge to normalise an expression, based on other information. One cannot determine precisely which "2 July" is referred to without a contextual clue of the year. These clues may be from the document creation time, or from a recently specified absolute temporal expression which sets reference time (see Sect. 6.3); failing that, the information again has to come from relations between temporal expressions.

2.3.3.2 Evaluating Temporal Expression Annotation

Precision and recall are suitable for evaluating temporal expression recognition (see Sect. 2.2.3.2). Temporal expressions can also be broken down into one of many classes and may be interpreted or even anchored to a calendar. To evaluate temporal expression typing, a simple "proportion correct" or accuracy metric works well. Interpretation and anchoring efforts can be compared verbatim to a gold standard to assess accuracy. One must also choose whether or not to allow equivalent matches to be considered as equal. For example, the TIMEX3 values `P1D` and `P24H` both correspond to a duration of a day and may be considered equivalent. However, if one prefers an annotation that matches the exact language used in a document, it may be argued that *"one day"* should only be given a value of `P1D` and that `P24H` is more representative of text like *"24 hours"*. Any timex evaluation needs to take a stance on these issues.

2.3.3.3 Timex Annotation Systems

Rule based systems are frequently employed in approaches to these tasks, because plenty of set phrases are used to describe time and they employ a simple grammar. In fact, one very successful approach to normalising week days is entirely rule-based [41]. This attribute of temporal expressions means that finite state grammars can be used for Timex recognition [13, 17]. In the case of the first paper, a rule-based system was completed by interleaving finite state grammars with named entity recognition, in order to enable temporal expressions in linguistic units, as opposed to lexical ones. This enables the identification of events and associations that are semantically present in a sentence but not immediately obvious from its construction.

Some systems, such as GUTime [26], rely heavily on a rule-based approach to spotting sequences of tokens, as there are many temporal expressions present in the English language that can be identified and anchored in this way. Named entity recognition (NER) has also been used to identify times in text [42].

Following MUC6, MUC7, TERN and ACE, TempEval-2 also included a task for temporal expression annotation. The entered systems and subsequent improvements have provided clear advantages over prior attempts in temporal expression annotation. Because timex annotation is not the primary focus of this book, only TempEval-2 and later experiments are described here.

In this task, rule-based, machine learning-based and hybrid systems all performed well at timex recognition. For English the timex extent recognition performance F-measure ranged from 0.26 to 0.86, with an average of 0.78. The best performance was with F1 of 0.86; seven systems reached F-measures of 0.84-0.86. This is promising, though by no means a solution to the timex recognition problem. Timex classification was performed best by a TIMEX2 transduction system with accuracy 0.98 [43], though all but two systems attempting timex classification reached at least 90 % accuracy.

Normalisation proved to be a substantially harder task, results ranging from 0.17 to 0.85. This task can involve complex reasoning and demands large and diverse amounts of training data [44]. The number of possible values is high, so giving default answers (e.g. most-common-class values) as a back-up is unlikely to be of any use.

Three systems in particular worked best at TempEval-2, though their strengths lie in different places. HeidelTime [45] is a modular rule-based system including a large ruleset; this enabled it to achieve top performance at timex normalisation, and has been used to power such applications as temporal information retrieval [46, 47]. However, rule-based approaches are likely to face diminishing returns as they attempt to raise recall through introduction of new rules [48]. TRIPS/TRIOS [49] and TIPSem-B [50] are both systems that use machine learning for timex recognition, with sophisticated feature sets. Using the TempEval-2 data released after the exercise, it has been shown to be possible for very simple feature sets to reproduce state-of-the-art timex recognition performance [51]. Normalisation remains a task that appears to requires a rule-driven solution, with promising new systems emerging [52].

2.4 Chapter Summary

This chapter has introduced the concepts of a timex and an event, and given formal definitions and annotation schemas for them, as well as describing the state of the art in their automatic annotation. We consider events and times as being anchored to a minimal representation in a document, typically a single word for events and a few words for temporal expressions (timexes). Conceptually, they are modelled as temporal *intervals*, having both a start and end instant, and holding for the period between. Events and times are the foundational building blocks of temporal discourse annotation, and both are considered as intervals whenever possible. The following chapter will cover the next step: temporal relations between intervals.

References

1. Moens, M., Steedman, M.: Temporal ontology and temporal reference. Comput. Linguist. **14**(2), 15–28 (1988)
2. Pan, F., Mulkar, R., Hobbs, J.: Learning event durations from event descriptions. In: Proceedings of the 21st International Conference on Computational Linguistics and the 44th annual meeting of the Association for Computational Linguistics, pp. 393–400. Association for Computational Linguistics (2006)
3. Gusev, A., Chambers, N., D.R., K., Khaitan, P., Bethard, S., Jurafsky, D.: Using query pattens to learn the duration of events. In: Proceedings of the 9th International Conference on Computational Semantics, pp. 145–154 (2011)
4. Mani, I., Schiffman, B.: Temporally anchoring and ordering events in news. Time and Event Recognition in Natural Language (2005)
5. Mani, I., Wellner, B., Verhagen, M., Pustejovsky, J.: Three approaches to learning TLINKS in TimeML. Tech. Rep. CS-07-268, Brandeis University, Waltham, MA, USA (2007)
6. Bethard, S., Martin, J., Klingenstein, S.: Timelines from Text: Identification of Syntactic Temporal Relations. In: Proceedings of the International Conference on Semantic Computing, pp. 11–18 (2007)
7. Pustejovsky, J.: The syntax of event structure. Cognition **41**(1–3), 47 (1991)
8. Cybulska, A., Vossen, P.: Historical event extraction from text. In: Proceedings of the 5th ACL-HLT Workshop on Language Technology for Cultural Heritage, Social Sciences, and Humanities, pp. 39–43. Association for Computational Linguistics (2011)
9. Ritter, A., Etzioni, O., Clark, S., et al.: Open domain event extraction from Twitter. In: Proceedings of the 18th ACM SIGKDD international conference on Knowledge discovery and data mining, pp. 1104–1112. ACM (2012)

10. Pustejovsky, J., Ingria, B., Sauri, R., Castano, J., Littman, J., Gaizauskas, R.: The Specification Language TimeML. In: The Language of Time: A Reader, pp. 545–557. Oxford University Press (2004)
11. Steedman, M.: Reference to past time, pp. 125–157. Speech, Place and Action (1982)
12. Vendler, Z.: Verbs and Times. The philosophical review. Ithaca, New York (1957)
13. Boguraev, B., Ando, R.: TimeML-compliant text analysis for temporal reasoning. In: Proceedings of International Joint Conference on Artificial Intelligence (IJCAI) (2005)
14. Verhagen, M., Saurí, R., Caselli, T., Pustejovsky, J.: SemEval-2010 task 13: TempEval-2. In: Proceedings of the 5th International Workshop on Semantic Evaluation, pp. 57–62. Association for Computational Linguistics (2010)
15. Buckland, M., Gey, F.: The relationship between recall and precision. J. Am. Soc. Inf. Sci. **45**(1), 12–19 (1994)
16. van Rijsbergen, C.: Information retrieval, 2nd ed. Butterworths (1979)
17. Ando, R.: Exploiting unannotated corpora for tagging and chunking. In: Proceedings of the ACL interactive poster and demonstration sessions. Association for Computational Linguistics (2004)
18. Saurí, R., Knippen, R., Verhagen, M., Pustejovsky, J.: Evita: a robust event recognizer for QA systems. In: Proceedings of the conference on Human Language Technology and Empirical Methods in Natural Language Processing, p. 707. Association for Computational Linguistics (2005)
19. Llorens, H.: A semantic approach to temporal information processing. Ph.D. thesis, University of Alicante (2011)
20. Derczynski, L., Strötgen, J., Maynard, D., Greenwood, M.A., Jung, M.: GATE-Time: Extraction of Temporal Expressions and Event. In: Proceedings of the 10th Language Resources and Evaluation Conference, pp. 3702–3708 (2016)
21. Hitzeman, J.: Semantic partition and the ambiguity of sentences containing temporal adverbials. Nat. Lang. Semant. **5**(2), 87–100 (1997)
22. Bestgen, Y., Vonk, W.: Temporal adverbials as segmentation markers in discourse comprehension. J. Mem. Lang. **42**(1), 74–87 (1999)
23. Cohen, D., Schwer, S.: Proximal deixis with calendar terms: Cross-linguistic patterns of temporal reference. ms. submitted in lingua (2012)
24. Strötgen, J., Gertz, M.: Domain-Sensitive Temporal Tagging. Morgan-Claypool (2016)
25. Hobbs, J.R., Pan, F.: An ontology of time for the semantic web. ACM Trans. Asian Lang. Inf. Process. (TALIP) **3**(1), 66–85 (2004)
26. Mani, I., Wilson, G.: Robust temporal processing of news. In: Proceedings of the 38th Annual Meeting on Association for Computational Linguistics, pp. 69–76. Association for Computational Linguistics (2000)
27. Ferro, L., Gerber, L., Mani, I., Sundheim, B., Wilson, G.: Tides 2005 standard for the annotation of temporal expressions. Tech. Rep. 03-1046, The MITRE Corporation (2005)
28. Ahn, D., Adafre, S., Rijke, M.: Towards task-based temporal extraction and recognition. In: Dagstuhl Seminar Proceedings, vol. 5151 (2005)
29. Han, B., Gates, D., Levin, L.: From language to time: A temporal expression anchorer. In: Proceedings of the 13th International Symposium on Temporal Representation and Reasoning (TIME) (2006)
30. Derczynski, L., Llorens, H., Saquete, E.: Massively increasing TIMEX3 resources: a transduction approach. In: Proceedings of the Language Resources and Evaluation Conference (2012)
31. Chinchor, N., Robinson, P.: MUC-7 named entity task definition. In: Proceedings of the 7th Message Understanding Conference (1997)
32. Allen, J.: Maintaining knowledge about temporal intervals. Commun. ACM **26**(11), 832–843 (1983)
33. Allen, J.F.: Towards a general theory of action and time. Artif. Intell. **23**(2), 123–154 (1984)
34. Derczynski, L., Llorens, H., UzZaman, N.: TimeML-strict: clarifying temporal annotation. arXiv preprint arXiv:1304.7289 (2013)

35. Pustejovsky, J., Knippen, R., Littman, J., Saurí, R.: Temporal and Event Information in Natural Language Text. Lang. Resour. Eval. **39**(2), 123–164 (2005)
36. Pustejovsky, J., Lee, K., Bunt, H., Romary, L.: ISO-TimeML: An International Standard for Semantic Annotation. In: Proceedings of the Seventh conference on International Language Resources and Evaluation (LREC'10) (2010)
37. Strötgen, J., Gertz, M.: Domain-sensitive temporal tagging. Synth. Lect. Hum. Lang. Technol. **9**(3), 1–151 (2016)
38. Androutsopoulos, I.: Temporal meaning representations in a natural language front-end. arXiv preprint cs/9906020 (1999)
39. Han, B.: Reasoning about a Temporal Scenario in Natural Language. In: Proceedings of the International Joint Conference on Artificial Intelligence (2009)
40. Mazur, P., Dale, R.: LTIMEX: Representing the Local Semantics of Temporal Expressions. In: Proceedings of the 1st International Workshop on Advances in Semantic Information Retrieval (ASIR), pp. 201–208 (2011)
41. Mazur, P., Dale, R.: Whats the date? High accuracy interpretation of weekday. In: 22nd International Conference on Computational Linguistics (Coling 2008), Manchester, UK, pp. 553–560 (2008)
42. Grover, C., Tobin, R., Alex, B., Byrne, K.: Edinburgh-LTG: TempEval-2 system description. In: Proceedings of the 5th International Workshop on Semantic Evaluation, pp. 333–336. Association for Computational Linguistics (2010)
43. Saquete, E.: TERSEO+T2T3 Transducer: a systems for recognizing and normalizing TIMEX3. In: Proceedings of the 5th International Workshop on Semantic Evaluation, pp. 317–320. Association for Computational Linguistics (2010)
44. Brucato, M., Derczynski, L., Llorens, H., Bontcheva, K., Jensen, C.S.: Recognising and interpreting named temporal expressions. In: Proc. RANLP, pp. 113–121 (2013)
45. Strötgen, J., Gertz, M.: HeidelTime: High quality rule-based extraction and normalization of temporal expressions. In: Proceedings of the 5th International Workshop on Semantic Evaluation, pp. 321–324. Association for Computational Linguistics (2010)
46. Strötgen, J.: Domain-sensitive temporal tagging for event-centric information retrieval. Ph.D. thesis, Heidelberg University (2015)
47. Derczynski, L., Strötgen, J., Campos, R., Alonso, O.: Time and information retrieval: Introduction to the special issue. Information Processing & Management (2015)
48. Bethard, S.: A synchronous context free grammar for time normalization. In: Proceedings of EMNLP, pp. 821–826. Association for Computational Linguistics (2013)
49. UzZaman, N., Allen, J.: TRIPS and TRIOS system for TempEval-2: Extracting temporal information from text. In: Proceedings of the 5th International Workshop on Semantic Evaluation, pp. 276–283. Association for Computational Linguistics (2010)
50. Llorens, H., Saquete, E., Navarro, B.: TIPSem (English and Spanish): Evaluating CRFs and Semantic Roles in TempEval-2. In: Proceedings of SemEval-2010, pp. 284–291. ACL (2010)
51. Llorens, H., Saquete, E., Navarro, B.: Syntax-motivated context windows of morpho-lexical features for recognizing time and event expressions in natural language. Nat. Lang. Process. Inf. Syst. **6716**, 295–299 (2011)
52. Llorens, H., Derczynski, L., Saquete, E., Gaizauskas, R.: TIMEN: An Open Temporal Expression Normalization Resource. In: Proceedings of the Language Resources and Evaluation Conference (2012)

Chapter 3
Temporal Relations

> The habit of looking to the future and thinking that the whole meaning of the present lies in what it will bring forth is a pernicious one. There can be no value in the whole unless there is value in the parts.
>
> *Conquest of Happiness*
> BERTRAND RUSSELL

3.1 Introduction

Having discussed timex and events in the previous chapter, we move on to discuss the temporal relations that exist between them. This chapter briefly describes temporal relations and surveys the state of the art in automatic temporal relation annotation. Extra attention is given to prior work on temporal relation typing. We will discover that temporal link typing remains a difficult problem, despite multiple sophisticated approaches. The overall picture highlights persistent difficulties in temporal relation typing and suggests that to understand how to temporally order events described in text, we need to draw upon multiple heterogeneous information sources.

Time can be described as a constantly progressing sequence of events. This sequential attribute is critical to the concept of a timeline, on which one may place events. Absolute locations upon the timeline are described using timexes. Conversely, event positions are not be absolute and sometimes can be temporally situated only in terms of their relation to other events or to timexes. This means that correctly identifying the temporal relations between pairs made up of events or timexes is critical to automatic processing of time in language.

In terms of information extraction, we are interested in either assigning an absolute temporal value to the start and end points of temporal entities, or describing these points in terms of other entities. It is helpful to have at least one value firmly anchored – normalised – to a timeline. If we have a specific distance between two events and the position of one has already been normalised, it is trivial to also normalise the other; for example, in "*John was born on the 24th April, 1942. His mother left the hospital nine days later.*", we have a "*born*" event which is already anchored and a "*left*" event which we can attach to 3nd May, 1942 with some inference.

L.R.A. Derczynski, *Automatically Ordering Events and Times in Text*,
Studies in Computational Intelligence 677, DOI 10.1007/978-3-319-47241-6_3

In cases where normalisation is not immediately possible, however, we may mark a relation between two events using a temporal link. This allows the representation of non-absolute temporal information. A network of events, times and relations help one to determine the temporal arrangement of events described in discourse.

While events and times are overt, the temporal relations that exist between them are abstract. Events and times in a text have lexicalised representations, but the ordering of them is not always made explicit. This contributes to the difficulty of temporal relation identification and typing.

The problem of reasoning about and of representing temporal information has been addressed in the fields of knowledge representation and artificial intelligence. Once a representation has been defined, we may formally describe certain temporal structures within a discourse and start to make inferences about temporal relations. Temporal relation types expressed in language do not necessarily match the classes available in an annotation schema. However, to perform automatic temporal relation extraction, it is important to decide a set of temporal relations. Part of the purpose of fixing this relation set is to aid inference; another is to provide a stable framework for human annotation.

In this chapter, we will first define the concept of temporal relations. This is followed by an exploration of different sets of temporal relation types applicable to linguistic annotation. After this, we discuss ways of annotating temporal relations over discourse, and the concepts of relation folding, temporal closure and temporal annotation as a graph are introduced. Next, the chapter introduces the general problem of automatic temporal relation annotation. This is followed by a literature review, coming up to the state of the art in automatic temporal relation typing. Finally, the chapter concludes with an analysis of the state of the art and the automatic relation typing problem.

3.2 Temporal Relation Types

Temporal algebras and logics allow one to deduce relationships between events based on their connection to other times and events, using a set of rules. These rules depend on the specific set of event relationship types and a set of relation types. Interval, point and semi-interval logics are all available. Building on STAG (Sheffield Temporal Annotation Guidelines, [1, 2]), TimeML (Sect. 2.3.2.1) defines its own set of interval relations, based on Allen's interval algebra [3]; point-based algebra can be useful for rapid reasoning; semi-interval reasoning relaxes the burden of specification required when both points of an interval need to be found, in order to avoid over-specification when working with events described by natural language and are discussed in Sect. 3.2.3.

For the context of this book, interval algebrae are considered to be those that define types of relation between intervals and a set of axioms for operating with these relations; an **interval** has a start and an end point. Some temporal logics use points instead of intervals. For interval logics, a point event may be represented by an interval whose start and end occur simultaneously; a **proper interval** is an interval where the end occurs after the start [4].

Temporal logics deal with reasoning about the relations that hold between intervals. Early examples of temporal logics include Prior's calculus for a modal tense logic calculus [5] and Bruce's model [6], which also includes axioms for event reasoning withing a temporal system.

This section first presents a few temporal interval algebrae, each with a specific purpose; finally, we will introduce the concept of temporal closure.

Applications of temporal logics can be found in multiple areas of computer science, including the verifying and testing time-sensitive parts of computer programs, in providing a temporal data representation for artificial intelligence systems and for representing temporal semantics in natural language processing. This section does not comprehensively discuss the full range of temporal logics, rather just those that deal with intervals and that have been previously applied to (or designed for) natural language processing. Other work has examined temporal logics in detail [7–9].

This section discusses some temporal interval algebras and their use in representing and reasoning over time as part of temporal information extraction. Firstly, there is a very minimal algebra, including just three relationship types. The limited number of potential relationship types makes it easier to visualise the relations between events and simpler to implement and troubleshoot problems that arise while reasoning. Secondly, we cover Allen's interval logic, which defines enough relations to cover all possible relations between a pair of temporal intervals. Finally is Freksa's logic based on semi-intervals, which tries to better capture and reason with the event relations pres in natural language discourse.

3.2.1 A Simple Temporal Logic

One can describe many basic relations between intervals using just three relations - BEFORE, INCLUDES and SIMULTANEOUS. If we encounter something such as *I washed after cleaning the sewer*, if events are denoted as E we can have simply reverse argument order to have $E_{cleaning}$ BEFORE E_{wash}. As part of a larger investigation into temporal reasoning on information found in discourse, [10] introduces a minimal logic based on three simple relations than only requires ten rules for temporal inference. The simplicity of this system makes it both easy to implement and easy to think about. However, the set of just three relations is small and the temporal relations expressed in natural languages can be more precisely represented using a wider set of temporal relation types. For example, if two intervals overlap but do not share any start or end points (such as winter in the northern hemisphere, which may begin in a November, and a calendar year), neither before, includes or simultaneous is precise enough to describe their temporal relation.

3.2.2 Temporal Interval Logic

Allen's interval logic [3] describes a set of temporal relations that may exist between any event pair. Allen introduces the concept of events (represented as intervals) as nodes in a graph, where the edges connecting nodes represent a relationship between two intervals. Where it is not clear that a single type of relation should exist between a pair of events, a disjunction of all possible relationship types is used to label the connection edge. Further, Allen provides an algorithm for deducing relationships between previously unconnected nodes.

The relations are listed in Table 3.1. Each of these gives a specific configuration of interval start and end points. Based on this, a transitivity table is provided for inferring new relations between intervals that hold common events. A full transitivity table is given in Table A.9.

A story typically describes more than one event, with some temporal ordering. Example 4 describes two events, setting out (E1) and living happily (E2).

Example 4 Little Red Riding Hood set out to town. She lived happily ever after.

The temporal link here is that she lived happily after setting out, signalled by both the textual order and also the use of the word *after*. Now, we can define a temporal link that says E2 AFTER E1 and label it L1.

It is improper to adventure without a cloak; perhaps we could introduce a new sentence in our text. See Example 5.

Table 3.1 Allen's temporal interval relations

Relation	Explanation of A-relation-B
BEFORE	Where A finishes before B starts
AFTER	Where A starts after B ends
DURING	Where A starts and ends while B is ongoing
CONTAINS	Inverse of DURING
OVERLAPS	Where A starts before B and ends during B
OVERLAPPED- BY	Inverse of OVERLAPS
MEETS	Where A ends at the point B begins
MET- BY	Inverse of MEETS
STARTS	Where A and B share their start point, but A ends before B does
STARTED- BY	As starts, but B ends first
FINISHES	Where A and B share their end point, but A begins later (and is thus shorter)
FINISHED- BY	As finishes, but B is the shorter/younger interval
EQUAL	Where A and B start and end at the same time

Example 5 Little Red Riding Hood set out to town. She put on her cape before leaving. She lived happily ever after.

This suggests a new dressing event, E3, signified by *putting on*. We also know the link between our new event and E1, setting out; E3 BEFORE E1. We'll call this L2. The story can now be represented by 3-node graph (events E1, E2 and E3), with two labelled edges (L1 and L2).

- E1: setting out
- E2: living happily
- E3: put on cape
- L1: E2 AFTER E1
- L2: E3 BEFORE E1

A visual representation of the temporal graph of these events and links is given in Fig. 3.1. This current graph leaves the relation between E3 and E2 unspecified. Narrative convention and human intuition tell us that we should use a linear model of time and suggest that anything that happens before the girl sets out must also happen before her living happily ever after. In this case, we can formally describe that knowledge with rules:

$\forall x, y : x \text{ AFTER } y \rightarrow y \text{ BEFORE } x$

$\forall x, y, z : x \text{ BEFORE } y, y \text{ BEFORE } z \rightarrow x \text{ BEFORE } z$

Thus, Little Red Riding Hood puts on her cape before living happily ever after and we can now introduce L3 as E3 BEFORE E2, completing the graph. This also describes BEFORE as a transitive relation.

Allen's logic was considered exciting because it was implementable at the time, unlike other temporal logics (e.g. [11]), and was also expressive; it has since been adopted by logicians, the verification and testing community and those interested in time in language. For a further review of temporal interval logics, one should see [8] and [12].

3.2.3 *Reasoning with Semi-intervals*

Temporal interval logic is not perfect. Determining consistency in any but the smallest scenarios quickly becomes intractable and is NP-hard [13, 14]. Problems arise when dealing with instantaneous events (e.g. "improper" intervals – Sect. 2.3); inconsistencies appear when events are allowed to have a duration of zero and the system is explicitly not structured to deal with these [15]. Semi-intervals are intervals where only one bound needs to be described (e.g. the start point or end point). It is contended that such relaxed definitions, when compared to fully-described intervals, can better represent the relations expressed in natural language. In this section, we discuss the shortcomings of temporal interval algebra and introduce a system for reasoning with semi-intervals.

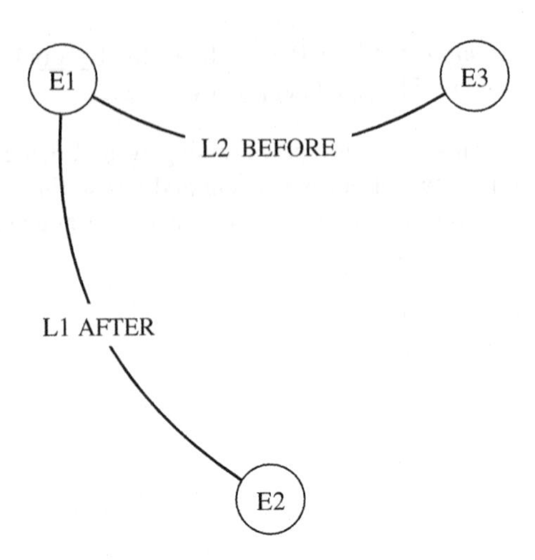

Fig. 3.1 Temporal graph of a simple story

Some common relation typing tasks are difficult to perform with interval relations. For example, newswire articles usually have a document creation time (DCT) or a publication date, which appears in document metadata and as a timex in the main body of discourse. They often contain at least a few events whose initiation is described in the past tense. In these cases, it is hard to determine whether an event's final bound stops at or continues past DCT, especially for states.

Example 6 contains an excerpt from a news report, uttered mid-way through a day. The timex *Today* has a specific meaning of a 24-h period. The start of the *control* event is unclear, but contextually we might assume that it begins before *Today*. Regardless of the arrangements of starting points of these two intervals, which could perhaps be discovered with further investigation, the arrangements of the endpoints of *Today* and *control* are unknowable at the time of utterance. Control could be relinquished before the day is over, at the precise end of the day, or later. This uncertainty makes it difficult to assign a relation from Allen's set to the two intervals. Without knowledge about the endpoints of these intervals, we can only say that the time-event relationship is one of *Today* $\langle overlapinverse, finishes, during \rangle$ *control.*

Example 6 Today, rebels still control the airfield and surrounding area.

To this end, [16] suggests a temporal algebra targeted at those dealing with natural language. It builds upon previous seminal work on logics that handle the uncertainties of time as described in language [17]. As long as we know that intervals begin before they end, we can start to describe relations between semi-intervals as disjunctions of Allen relations. It is quickly observed that particular Allen relations occur together, when dealing with incomplete knowledge about events. Freksa summarises these, defining terms for conceptual neighbours – "two relations between pairs of events are **conceptual neighbours** if they can be directly transformed into one another by continuously deforming (i.e. shortening, lengthening, moving) the events (in a topological sense)". For example, BEFORE and MEETS neighbour, as one can change the relation between two events from one of these to the other by adjusting the endpoint of the interval that starts earliest. We then also have **conceptual neighbourhoods**, which are sequences of relations which are conceptual neighbours.

Freksa's system tackles uncertainty about knowledge linking two events and allows us to capture information from text that may not describe all intervals completely. Using groups of relations that commonly co-occur during inference, Freksa describes a temporal algebra, labelling certain groups of Allen relations as relations in their own right. The algebra specifies a transitivity table. The table is based on commonly co-occurring groups of relations.

For example, from Freksa's set, the relation A `older` B applies whenever A's start point happens before B's start point; no attention is paid to their endpoints and so any of A [BEFORE, IBEFORE, ENDED_BY, INCLUDES] B apply. From this example at least one instance in English where a semi-interval logic would be useful is immediately clear. Further examples are provided in Freksa's paper. Additionally, Sect. 6.4.2 investigates semi-interval logic in the context of tense-based temporal relation typing.

3.2.4 Point-Based Reasoning

As their name suggests, point-based temporal logics work only with the ordering of individual points and do not cater for the concept of an interval. They are less prone to the over-specification problem that full interval algebras have (see above). It is possible to decompose intervals to their beginning and end points. Only equality and precedence operators are needed to described binary relations between these points. Point-based algebrae can be very fast to process, a feature which tools such as SputLink [18] and CAVaT [19] exploit. They also better lend themselves to graph-based reasoning about temporal structures in text [20]. However, it is more complicated for humans to annotate using points instead of intervals and the semantics of temporal relations in text are better represented with interval or semi-interval labels. Because of these reasons and because temporal annotation is already a difficult and exhausting task for human annotators, point-based reasoning and temporal logics are generally restricted to the domain of fully automated reasoning [8].

3.2.5 Summary

We have outlined the requirements for temporal logic in the context of language and detailed examples; a simple 3-relation logic, Allen's interval logic, Freksa's semi-interval logic, and point-based reasoning. In the next section, we will see how using these logics with an existing document can tell us about temporal links that have not yet been annotated.

Table 3.2 TimeML temporal relations

Relation	Explanation of A-relation-B
BEFORE	A finishes before B starts
AFTER	A starts after B ends
INCLUDES	A start before and finishes after B
IS_INCLUDED	A happens between B's start and finish
DURING	A occurs within duration B
DURING_INV	A is a duration in which B occurs
SIMULTANEOUS	A and B happen at the same time
IAFTER	A happens immediately after B
IBEFORE	A happens immediately before B
IDENTITY	A and B are the same event/time
BEGINS	A starts at the same time as B, but finishes first
ENDS	A starts after B, but they finish at the same time
BEGUN_BY	A starts at the same time as B, but goes on for longer
ENDED_BY	A starts before B, but they finish at the same time

3.3 Temporal Relation Annotation

The work in this book primarily concerns temporal relation annotation using intervals, as opposed to points or semi-intervals. This section is about turning the abstract idea of temporal ordering into something well-defined that we can reason with directly – the process of annotation.

Temporal relations obtain between two endpoints. They describe the natural of a temporal relation between those endpoints. Those endpoints my be either times or events, and needn't be of the same type. Therefore, a temporal relation annotation must at the minimum specify two endpoints and a relation (or label describing the relation) that exists from the first to the second. Optionally, additional information may be included, such as pointers to phrases that help characterise the relation.

There are three sets of temporal relations commonly used for linguistic annotation: Allen's original set (Table 3.1), the TimeML interval relations (Table 3.2), and the TempEval-1 and TempEval-2 simplified set (Table 3.3).

The TimeML relations are intended to be interpreted slightly less strictly than the Allen set. As language is imprecise and there is often some uncertainty around the precise location of endpoints, a little variance is permitted; actual events need not

Table 3.3 The relation set used in TempEval and TempEval-2

Relation	Explanation of A-relation-B
BEFORE	Where A finishes before B starts
AFTER	Where A starts after B ends
OVERLAP	Where any parts of A and B co-occur
BEFORE- OR- OVERLAP	A disjunction of BEFORE and OVERLAP
OVERLAP- OR- AFTER	A disjunction of OVERLAP and AFTER
VAGUE	For completely underspecified relations

start and end at the exact same (e.g.) millisecond[1] – instead, interpretation is left to the annotator.

TimeML describes realis, non-aspectual temporal relations using the **TLINK** element. The TLINK element's **relType** attribute's value is that of the temporal relation's type.

3.3.1 Relation Folding

Many of the relations used in both TimeML and Allen's interval algebra have an inverse relation, which they can be mapped on to by simply substituting the relation type and switching over the argument order. For example, BEFORE(monday, tuesday) is equivalent to AFTER(tuesday, monday). Automatic classification is easier with a smaller number of classes. We can simplify the task of classifying temporal relations by reducing the set of relation types used.

The procedure of removing inverse relations requires the definition of a set of mappings from relations with their complements. Using this, one removes inverse relationship types by changing them to their original form and flipping argument order. We have named this procedure **folding**.

Various relation folding mappings are available. MITRE specifies one (for example, those used by [21]) and there are mappings to the simple SIMULTANEOUS/BEFORE/INCLUDES relations specified by [10]. To be able to accurately reproduce results, one requires a dataset where the set of relation types has been reduced (folded) in the same way.

Although it may at first seem that folding relations in a document will alter the distribution of relationship classes, it must be pointed out that the exact balance between BEFORE and AFTER relations – indeed between any relation and its inverse – is entirely arbitrary and down to the annotator's personal preference. Folding in

[1] Although scale plays a part here; for some events, starting within the same week or even millennium can be considered synchronous, for others, picoseconds can be considered apart. The final choice is left to the annotator, who should interpret discourse accordingly.

fact removes any influence that annotator preference may have and presents data in a uniform manner.

Based on Table 3.1 from [21], MITRE have opted for the following mappings: (an asterisk indicates that the arguments should be reversed as part of the relation type change)

- IAFTER → IBEFORE*
- BEGUN_BY → BEGINS*
- ENDED_BY → ENDS*
- IS_INCLUDED → INCLUDES*
- AFTER → BEFORE*
- IDENTITY → SIMULTANEOUS
- DURING → INCLUDES*
- DURING_INV → INCLUDES

This gives us a smaller set of six relations, from the original fourteen. The mapping suggested by [10], from [13], is reproduced in the same format here:

- AFTER → BEFORE*
- IS_INCLUDED → INCLUDES*
- IDENTITY → SIMULTANEOUS
- DURING → INCLUDES*
- IBEFORE → BEFORE
- IAFTER → BEFORE*
- BEGINS → INCLUDES*
- ENDS → INCLUDES*
- BEGUN_BY → INCLUDES
- ENDED_BY → INCLUDES

There has been ambiguity over how best to fold DURING relations. After some discussion [22], the TimeML DURING relation can be said to specify a relation between two proper intervals that share the same start and endpoints (cf. "for the duration of") and that DURING is formally equivalent to SIMULTANEOUS; as SIMULTANEOUS is

Table 3.4 Relation folding mappings used in this book

Original relation	Folded to
AFTER	BEFORE*
IS_INCLUDED	INCLUDES*
IAFTER	IBEFORE*
BEGUN_BY	BEGINS*
ENDED_BY	ENDS*
DURING_INV	SIMULTANEOUS
DURING	SIMULTANEOUS
IDENTITY	SIMULTANEOUS

the inverse of itself, nothing unusual need be done for DURING_INV, which resolves to the same type. After this clarification, the fold used in experiments detailed by the rest of this document is shown in Table 3.4.

The effect that folding has on the distribution of link types in the TimeBank corpus can be observed by comparing Tables 3.5 and 3.6.

Table 3.5 Distribution of TLINK relation types in TimeBank 1.2

Relationship type	Count	Percentage (%)
AFTER	897	14.0
BEFORE	1408	21.9
BEGINS	61	1.0
BEGUN_BY	70	1.1
DURING	302	4.7
DURING_INV	1	0.0
ENDED_BY	177	2.8
ENDS	76	1.2
IAFTER	39	0.6
IBEFORE	34	0.5
IDENTITY	743	11.6
INCLUDES	582	9.1
IS_INCLUDED	1357	21.1
SIMULTANEOUS	671	10.5
Total	6418	

Table 3.6 Distribution of relation types over TimeBank 1.2, as per Table 3.5 and folded using the mappings in Table 3.4

Relationship type	Unclosed		Closed	
	Count	Percentage (%)	Count	Percentage (%)
BEFORE	2305	35.9	22033	73.2
BEGINS	131	2.0	226	0.8
ENDS	253	3.9	479	1.6
IBEFORE	73	1.1	169	0.6
INCLUDES	1939	30.2	4368	14.5
SIMULTANEOUS	1717	26.8	2822	9.4
Total	6418		30097	

3.3.1.1 Problems with Folding

While folding reduces the number of possible relation classes and increases the amount of training data available in each class, it introduces some system implementation issues. In controlled evaluation exercises, it is possible to reverse the order of arguments in the evaluation set such that the set only contains relations that the classifier has seen before from folded training data. However, this is not possible in cases where the relation type is never known. One does not have control over the argument order of unlabelled examples that are to labeled. If for example we have removed all AFTER relations from our training data by swapping their arguments and changing the relation to BEFORE, when faced with the previously-unseen relation of (e.g.) "C AFTER D", the classifier will not be able to assign the correct label. One solution is to attempt to classify the intervals twice – A rel B as well as B rel A – and use classifier confidence or the addition of an "unknown" relation type to signify which of the reduced label set should be applied with which arrangement.

Another approach for building applications that can cope with non-synthetic data is as follows. Maintain the normal set of relations and increase training data size by using folding to create a new training instance (instead of folding to alter a training instance) and add that to the set. That is, if we have a training example "A AFTER B", we automatically add an example of "B BEFORE A" and leave both examples in the training set. This technique can be called relation **doubling**. When performing doubling in this manner, it is even more important to partition training and testing data at document and not example level.

In summary: classifiers trained on folded data may not be able to cope with real-world data; classifiers learning from data created by doubling do not have such a disadvantage; folding works by simplifying the training data; doubling works by increasing its volume.

For the sake of comparability, the work in this book is uses training data with folded relations. Investigation of temporal relation doubling as a replacement for temporal relation folding is left for future work.

3.3.2 Temporal Closure

Humans tend to first classify the links where they find the type most obvious, de-prioritising other more tenuous or remote links [23]. Thus, out of all possible links between each event and temporal expression, usually only a subset of links are classified by a human annotator. It is possible, however, to determine a canonical version of the temporal structure of a document.

Smaller datasets are problematic for automated approaches to relation typing because they may not contain sufficient information to form generalisations about relations. Further, temporally annotating documents in order to enlarge datasets is a complex and costly procedure. Therefore, any automated aids to increasing the amount of temporal relations annotated are welcome. Fortunately, it is usually possible to automatically perform some inference over an incomplete annotation, labelling extra edges with relations and thus reducing data sparsity. One may use a temporal algebra to infer relationship types.

Let times and events be nodes on a temporal graph and edges in the graph represent relations between them. Given a partially connected temporal graph (for example, a human temporal annotation of a document), one can iteratively label previously unlabelled edges using an algebra's inference rules. When no more unlabelled edges can be labelled, the resulting graph represents the **temporal closure**. This graph explicitly conveys the maximum amount of information that one is able to deduce from a partial annotation. Once the maximum number of interval pairs have been linked in this manner, we are said to have computed the **temporal closure** of a document. For an example, see Fig. 3.1. Graph-based representations lead to sophisticated reasoning [20] and evaluation measures (Sect. 3.4.4.3).

There is often more than one way of temporally annotating a document's temporal structure. Because there is often more than one way to annotate a document that can be computed to the same temporal closure, when comparing documents, the closure is used rather than the original annotation. Closure also provides extra training examples for supervised learning, which has been explored by many authors, particularly investigated by [24] (see Sect. 3.4.1). We fully investigate comparison of temporal annotations in Sect. 3.4.4.

3.3.3 *Open Temporal Relation Annotation Problems*

Within temporal relation annotation, there remain open problems in a number of areas. This book contributes towards the solution of one – temporal relation typing. Others are detailed here.

Temporal Relation Identification

This is the task of determining which pairs of events or timexes should be linked. While one may link almost every time and event annotation in a document by means of inference (perhaps through closure), is this the best option? Adding structure to the relation identification task often leaves out some links that are otherwise clear to readers. For example, the TempEval exercises focus on intra-sentence links between the head event and other events, and then on head events between adjacent sentences – but this says nothing about the relation between non-head events in the same sentence. Determining a definition of what constitutes a temporal relation and then finding these in text remain open.

Modality

The majority of research has focused on links between events and times in the same modality and in the same frame of reference. Dealing with modals seems important; they occur frequently, and indeed there is a strong argument that the future tense is entirely modal. The problem of temporal annotation between non-concrete modalities is open.

Annotation Completeness

How do we know that we've finished annotating? Even given oracles for event annotation, timex annotation, and temporal relation identification and typing, there exists no firm description of what constitutes a complete annotation. Is it when every event and timex is connected? Is it just when those links based upon explicit temporal words and inflections in the text have been annotated? Neither TimeML nor other temporal relation schemas tackle the problem of annotation completeness. As temporal relation annotation in particular is a difficult and time-consuming task, it would be very helpful to establish at least recommended minimum and maximum bounds for relation annotation.

For a really good guide to annotation in general, I recommend "Natural language annotation for machine learning" [25].

3.4 Automatic Temporal Relation Typing

Over the past decade or so, there have been many machine learning approaches to temporal relation typing – the task of determining the relative order (or relation type) between two temporal intervals (which are times or events). Most of these approaches have focused on using a set of relations derived from the 13 labels proposed by Allen (Table 3.1) or a reduced set thereof (e.g. TempEval relations, Table 3.3). The most commonly used datasets are TimeBank and TempEval-2 (Section A.2).

Generally, earlier relation typing systems are accurate in around 60 % of cases and more recent systems reach about 70 % accuracy. This level is only ever exceeded in cases where a subset of all temporal links is examined; never for the general problem.

This chapter describing related work first summarises some concepts particularly useful to temporal relation typing (Sect. 3.4). After this, a set of previous approaches are described, in terms of their dataset, features and performance (Sect. 3.5). The progress in the field so far is then summarised and an analysis presented (Sect. 3.5.6).

3.4.1 Closure for Training Data

In order to provide extra training data, temporal closure [26] can be performed over human-annotated data. This provides a varying number of additional examples, depending on the completeness of the initial annotation (perhaps symptomatic of the lack of a formal definition describing how much should be annotated) and also the text itself.[2]

[2]Examined in greater detail in Sect. 3.3.2.

3.4.2 *Global Constraints*

In linked groups, temporal relations co-constrain. For example, given:

Example 7

A BEFORE B
B BEFORE C

The set of valid types for an A–C relation is constrained. It is important that automatic labellers take this knowledge into account. The production of an overall inconsistent annotation is a simple thing to check for. In all but the simplest of documents, global co-constraint violates the independence of training examples. In order to preserve separation between training and test data, [24] propose only allowing document-level splits in data.

3.4.2.1 Event Sequence Resources

As we annotate text, it becomes possible to build some discourse-independent record of common event relations. This is essentially a restricted model of world knowledge. For example, we might often see that *travel* happens before *arrive*, or that *sunrise* is included in *the day*. Such records could be used to aid future annotation of unlabelled temporal relation data.

VerbOcean

One such resource that specifies a simple relation between token pairs is VerbOcean [27, 28]. The data comes from mining Google results using templates [29] and then establishing mutual information between mined verb pairs. Different relation types each have their own set of templates. The relations that are useful in temporal information extraction are `[happens-before]` and `[can-result-in]`, reflecting causation and enablement.

Narrative Chains

Chambers and Jurafsky [30] suggest a way of building event chains. These look for common actors in events (either as subject or object) and catalogue the events that the actor participates in. Actors do not need to be people in this context. Event chains are provided in a number of different story types. An example is given where a criminal robs, and then is arrested, and is tried; this sees the “criminal” actor fulfil multiple roles. When a particular chain of events can be seen to occur in the same sequence (with similar actors) over many documents, we can have higher confidence in its accuracy. While this work does not suggest any kind of temporal ordering, it is easy to see how one can build catalogues of temporally sequential stories, which may later be of use when ordering events.

3.4.3 Task Description

The task of determining which times/events to relate is "temporal relation identification". The task of determining the type of relation that holds between a given timex or event pair is "temporal relation typing". This chapter concerns the temporal relation typing task: that is, of assigning on of a set of relation types to a given interval pair, where an interval may be an event or timex.

Consider the sentence in Example 8.

Example 8 The president's son met$_e$ with Sununu last week$_t$.

It contains an event e and timex t. We are told by an external source, e.g. our annotators, that has already performed temporal relation identification, that e and t are temporally related. The task at hand is to choose a relation type from a set of options that best describes the temporal relation between e and t. A list of these options in TimeML is in Table 3.2.

In this scenario, the *met e* seems to occur in its entirety at some time between the beginning and end of *last week t*. So, the suitable relation type is inverse inclusion; that is to say, e IS_INCLUDED t. Or, the other way round, *last week* INCLUDES *met*.

3.4.4 Evaluation

In many tasks related to temporal processing of text, there is a need to compare annotations. One may want to compare two human annotations, or measure how favourably an automatic annotation compares to an existing gold standard. Developing an automated temporal information extraction tool in any kind of scientific way requires formal evaluation. Comparing two human annotations will give values for inter-annotator agreement (used as a rough cap for automatic annotation performance) and the ability to evaluate automatic systems is essential.

Human annotation of temporal relations is difficult [10, 31]. This is sometimes caused by a lack of context during annotation. For example, some systems show only two event sentences, omitting surrounding discourse which may contain clues [32]. Humans, for example, have trouble distinguishing some relations such as IS_INCLUDED and DURING [33]. The temporal relation annotation task is complex enough to have a large number of idiosyncratic difficulties, which we can only identify through annotation comparison.

In the rest of this section, we introduce general issues with temporal relation evaluation and then discuss the application of traditional precision and recall measures to this task, as well as two graph-based methods for comparing temporally-annotated documents.

3.4.4.1 General Issues

Temporal relation annotation evaluation involves the assessment of relation type assignments between an agreed set of nodes. Because of the complex nature of the interactions between relations that share nodes, the following issues need to be taken into consideration when evaluating temporal relation typings.

Firstly, with most relation sets there is more than one way of annotating a single relation between two events or times. One may say "A BEFORE B" or "B AFTER A", both describing the same temporal relation between A and B.

Secondly, the transitive, commutative and co-constraining nature of temporal relations in a network mean that there are many different ways of representing the same information [10, 26, 34] in the form of a temporal closure. As a result, missing links are not always a problem, as long as the information required to infer them is present somewhere in a document. As a general approach, one should only evaluate over the closure of a document's annotation.

Finally, when evaluating it is important to take account of which document an instance of a relation comes from. Mutual co-constraint means that relations within a single document or temporal graph are not independent. When partitioning data into training and test sets, one must be careful to split at document level; that is, all links from any document should be in the same set. When performing cross-validation, all of each document's links should be found only in one single fold [24].

3.4.4.2 Precision and Recall

Annotations can be compared in different ways. When evaluating automated TIMEX identification or relation classification against a gold standard, we can measure precision and recall. For example, one can use these metrics to describe the amount of TLINKs correctly found in a candidate annotation versus a reference annotation. TimeBank is often used as a gold standard for training and evaluation of systems using TimeML. Evaluating TIMEX normalisation needs a different measure, as there are varying degrees of correctness available; one has to take granularity into account, as well as potentially overlapping answer intervals, which should not automatically be granted zero score.

Sometimes important links will be missed by annotators; sometimes multiple unclosed annotations of the same closed graph can differ. The latter can be compensated for by only comparing closures; in fact, precision and recall should only be measured between closed graphs, otherwise there is misleading ambiguity between different representations of the same information. Measuring the presence of relations only affects recall; unlabelled edges are equivalent to missing information, as opposed to incorrect information.

3.4.4.3 Graph-Based Evaluation

While precision and recall provide an indication of the closeness of two annotations, they are imperfect in the context of temporal annotation. Flaws exist in relation type matching and evaluating interval boundary point assignment. For relations, some temporal link types are more closely related than others. If we guess INCLUDES when the real answer is ENDED_BY, we have done much better than if we guess BEFORE. For intervals, working at interval level requires both endpoints to be correct before awarding a full entity match. However, it is rational to issue a partial reward if one endpoint has been found correctly, when compared to cases where neither are correct. Precision and recall based systems cannot directly cater for these features of these problems. This section discusses a graph-based evaluation metric that attempts to address these issues.

As mentioned in the chapter introduction above, a discourse's temporal information can be imagined as a graph (see Sect. 3.2.2). Temporal closure of the graph can be computed, leading to a more consistent representation of the annotated data [3, 10, 13]. It is possible to measure agreement between graphs [32].

Not all relations have the same importance; some entail more information – some may lie on something akin to a critical path [35], and conversely some may only be dead ends that do not affect the rest of the graph. Resolving certain relations provides more information than others. Thus, a metric that rewards the labelling of the most important edges is required.

One can use a graph algebra to build a metric for graph similarity. One method of achieving this, proposed by Tannier and Muller [34], involves the following steps:

- Graphs between events are converted into graphs between points
- Each event is split into a beginning and end point
- Only equality (=) and precedence (<) relations are needed
- Two nodes linked by equality relations are merged

This produces an acyclic directed graph, of arcs which represent precedence relations, and nodes that represent collections of temporally simultaneous points. An edge between time points x and y implies that x is equal to or less than y. The transitive reduction of a directed acyclic graph, which is unique, is calculated. After this, Allen relations are converted into '=' and '<' (equality and precedence) relations between endpoints. At this point, we have a linear directed graph, with one or more points (each representing an interval start or end point) at each node. From the directed graph, multiple candidate graphs can be compared by the number of manoeuvers required to reach one graph from the other, in a similar fashion to establishing a Levenshtein edit distance [36].

Manoeuvers are of two types. A **split** is where a node is broken and a **merge** is the addition of a point to a node.

The similarity between graphs is measured based on the number of merges and splits required to transform them, over the total number of relations. One can then calculate a revised version of 'temporal' recall and precision, based on features in the graphs. Graph value, representing the size and complexity of a graph, is key to these measures. It is also possible to evaluate graphs that include temporal relations of the form 'before or equal" ('after or equal" is reduced to this form by reversing arguments). Half-splits and half-merges can be introduced, with an initial weighting of 0.5 for the move, where a half-split would be the removal of a point related with such a disjunction.

To see how useful this evaluation metric is, its authors used it to examine graphs where selections of temporal relations had been removed from minimal graphs and a linear decrease in the standard recall measure was observed (as expected). However, while recall harshly penalises graphs that lack some critical information, this metric still rewards the remaining partial information, leading to a convex graph curve, which can be seen in Fig. 16 of [34]. Thus, this measure provides an intuitive metric for temporal annotation comparison which offers partial rewards for partially correct information, unlike precision and recall measures.

Although an improvement upon earlier metrics, graph-based evaluation is used little in the literature and so experiments measured using can be difficult to compare to previous work; e.g. [37].

3.4.4.4 TempEval

The TempEval semantic annotation evaluation exercises are shared tasks focusing primarily on temporal relation annotation. They have also served to advance the state of the art in temporal annotation [38]. TempEval and TempEval-2 both use a simplified set of relations and a purpose-created corpus. Systems in TempEval-2 [39] showed some incremental relation typing performance improvements over the previous exercise. While the first TempEval focused on the temporal relation typing task, TempEval-2 added event and timex annotation, and TempEval-3 [40] also required participants to perform temporal relation identification. These three establishing evaluation challenges led to us seeing a proliferation of temporal evaluation challenges; in 2015 we saw not one but four different temporal shared challenges at SemEval, covering cross-document coreference and ordering, question answering, clinical data, and document data [41–44]. Clearly temporal semantic annotation is an area full of tough and fundamental challenges.

TempEval has generally contributed extra data and served to advance the state of the art, not only by stimulating research as many different sites contribute systems but also by providing empirical, comparable results for many different approaches to temporal annotation.

3.5 Prior Relation Annotation Approaches

This section presents an overview of automatic temporal relation typing efforts. It aims to be comprehensive, especially to include work done after the introduction of TimeML. It is broken into the discussion of machine learning-based systems, rule-based systems and hybrid systems. Several techniques for boosting training data size and feature effectiveness are discussed. Finally, an analysis is presented in which successful parts of an approach are identified and future work is outlined.

3.5.1 Feature and Classifier Engineering

Many approaches have relied on using example relations to train a classifier, i.e. are supervised learning approaches. These relations are represented as a vector of features. It is critical to select the right features and classifier, and these have been topics of many prior approaches.

Machine learning approaches do not require an intimate and accurate human understanding of all linguistic relations within a document. Rather, a classifier learns rules or models from training data and uses these to attempt to predict the label of future relations given their feature vector representation.

Classifier performance generally improves as more training data becomes available. This has the benefit of being able to directly boost performance through data collection. However, insufficient training data can lead to poor performance, and in the context of temporal annotation, collecting more data is expensive. In the case of temporal information extraction, relatively small amounts of ground truth data are available.

With linguistic datasets, it is important to choose a classifier that can resist some noise in its training data. Natural language is robust and many utterances can be understood despite some minor mangling. Further, the diverse range of words that may be used in any situation are prone to inducing overfitting if not handled correctly. We shall see this later, in for example Sect. 5.6.3.5.

One of the earliest approaches [45], shortly after the release of TimeBank 1.1 (which included timex, event and relation annotations), attempted to both determine which intervals to link (the relation identification task) and then also to determine the nature of the TimeML relation between detected pairs (the relation typing task). It used an RRM classifier [46] to jointly detect and label TLINKs based on features derived from a finite state parser. These were based on the gold-standard event and timex annotations in that corpus. Only event-timex links were considered. A proximity threshold for intervals classified as being temporally linked or not was set. This proximity threshold was varied in an attempt to discover its impact on the complexity of the task. The baseline for pairing was that only if an event and timex were the closest of their kind to each other would a link be said to exist, and the baseline for typing was most-common-class (IS_INCLUDED). Features are based on part-of-speech tags, word shapes, syntactic chunk information and n-grams.

Only looking for TLINK argument pairs within 4 tokens provided the strongest results at the pairing task (F-measure 81.8). When the authors have to both find the TLINK and then assign a relationship type (a harder task than we address in this book), the F-measure dropped to 58.8. This indicates a typing accuracy of around 70 % in this small subset of TLINKs. Adding FS grammar information (see also

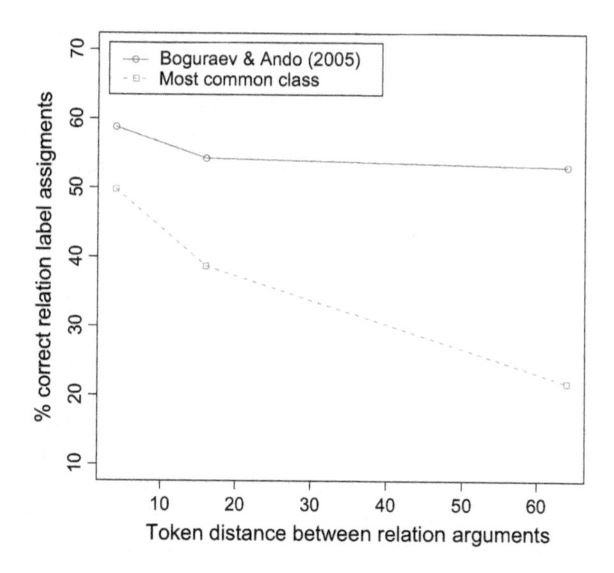

Fig. 3.2 TLINK relation type assignment difficulty increases with the distance between link arguments

Sect. 2.3.3.1) to the feature set consistently provides a small absolute performance boost (0.7–1.8 %). They found that automatic detection and typing was easier for relations between intervals the closer that they were in discourse, reaching 58.8 % accuracy on the joint TLINK-finding/relationship assignment task for interval pairs within four tokens of each other (which accounts for 12 % of TimeBank's relations). This accuracy decreased with larger token window sizes (see Fig. 3.2, which is derived from data tables in their paper). Considering EVENT/TIMEX3 pairings in the largest window size – 64 tokens – yields a low baseline performance of 21.8 %; the classifier improves on this to reach 53.1 % at this joint relation identification/typing task.

It is possible to determine the performance of [45]'s joint relation identification/ approach at just the relation typing task. Dividing joint pairing/typing performance by typing performance gives the typing accuracy over correctly identified relations. In this case, for 4-, 16- and 64-token windows respectively, TLINK typing using the features above including FS grammar information reached 71.9, 71.0 and 71.0 % accuracy respectively. These figures apply to event-timex links between intervals that appear relatively close to each other in discourse.

As part of TempEval 2007, [47] experimented with a range of classifiers and the basic event/timex attributes as features, attempting to gather information on which attributes were helpful in relation typing. Among other things, they found that tense and aspect features were of less use in event-timex relation typing than in event-event, and that SVM and K* classifiers performed best.

After the release of TimeBank v1.2, upon which the majority of recent temporal relation extraction work is based, [21] proposed a supervised learning approach to event-event and event-time relation typing, using the interval pairings specified in the corpus. This was refined and presented later [24] as an approach that provides a useful baseline for other supervised approaches, as it relied only upon information annotated with TimeML (e.g. no n-gram or syntactic features). The features used for each link were the text and TimeML element attributes of the intervals comprising the link, as well as a few simple Boolean features describing whether or not the tenses and aspects of both participants in an event-event relation were the same. The authors experimented with using temporal closure to increase the number of relations available (see Sect. 3.3.2).

The corpus used is a merging of a custom version of TimeBank [48] (v1.2a – not publicly available) and the Aquaint TimeML Corpus (ATC) [49]. Applying a maximum entropy classifier (from Carafe[3]) reaches an accuracy of 82.5 % when classifying event-to-time relations, better than the most-common-class baseline of 65.5 % (this class is the INCLUDES relation). Event-event relations were labelled with 59.7 % accuracy, which improved on the most-common-class baseline of 51.7 % (BEFORE). Other classifiers – namely SVM and naïve Bayes – performed similarly. As for using data from temporal closures of the annotations in the source corpus, event-time typing was better than baseline but overall worse (71.2 % accuracy, 51.3 % baseline) but event-event typing did worse than most-common-class baseline (51.1 % accuracy, 54.1 % baseline). Generally, classifiers trained on unclosed data performed better when predicting labels for TLINKs from unclosed data than did classifiers trained on closed data (at predicting TLINKs from closed data). This suggests that simply generating extra feature instances via temporal closure of source data data is not an effective method for learning better classifiers.

Later approaches have adopted the method used by [24] – that is, using a combined TimeBank/AQUAINT corpus plus the TimeML element attributes as features. Using support vector machines, [50] achieved performance gains in TimeML temporal relation typing using syntactic tree kernels. Their approach reached 80.04 % accuracy on event-time links in ATC using a polynomial composite kernel (compared to 82.47 % from [24]) and 67.03 % for event-event relations on the same (compared to 70.4 % from [51], detailed below).

Vasilakopoulos [52] use a K* approach to temporal relation typing. They determine the most useful features for the typing task and discard the least useful, as well as experimenting with new semantic features. This leads to strong performance on the earlier TimeBank 1.1 corpus.

3.5.2 Rule Engineering

As opposed to supervised machine-learning approaches, some approaches to automatic temporal relation typing use a human-engineered set of rules to determine how to assign a relation label. These rules are typically based on information about the relation and its arguments. These approaches can be simple and intuitive and quickly achieve above-baseline performance with a minimal ruleset. However, to reach com-

[3] Available at http://sourceforge.net/projects/carafe/.

petitive accuracy levels, the rule set generally becomes more complex and harder to understand.

Rule based approaches tend to be more fragile than generic learned approaches. Extrapolation can be a particularly difficult task, which can occur when coping with unseen data that does not match patterns previously seen. Further, performance is not dependant on the amount of training data, but instead the quality of the rule set. Therefore, one cannot directly turn extra data into better accuracy.

That said, there are still some rule-based approaches that have met with success. Initial work on the relation typing task was conducted by [53], using a rule-based technique to anchor events to times. This rule-based technique draws on principles from Reichenbach's model of tense and aspect [54]. They achieve an 84.6 % accuracy, though the work is hard to compare to later approaches based on TimeML because the relation set is simplified and the event and time definitions are not the same.

It is possible to add rules to a system which support incorrect decisions in some cases. Such rules will damage performance. However, including only high-performance rules becomes increasingly difficult as more rules are added to a system, and can constrain the scope of new rules to only cover a few cases. Kolya et al. [55] describe a rule-based approach that includes rules which have known contradictions in the training dataset. This approach has intentionally capped its maximum performance. Despite this, is it still able to achieve reasonable accuracy on its evaluation set.

The sentiment that neither rule-based nor statistical methods alone can satisfactorily solve a qualitatively described real-world problem is not a new one [56]. Hybrid approaches can overcome problems with both rule-based and machine learning-based options. Rule based systems have problems with rigidity and with their high construction cost; machine learning systems can quickly make inferences over data, but rely on having both accurate data and enough data. With a hybrid system one can incorporate rules to quickly achieve a base performance level and a machine learning component can "weight" rules to avoiding some of the fragility of complex rule bases. Further, one can quickly and simply prototype a machine learning system and then provide expert knowledge in the form of rules, allowing a rapid way of building new information into an automatic labeller. As a result, rule engineering has been used in combination with machine learning by many approaches to the relation typing task.

Kolya et al. [57] augment a CRF-based event-time relation typing system with a set of hand-crafted rules that encode observations about the dataset, leading to strong performance for event-event and event-time relation typing. In later work they take a similar approach [58], using event head information to achieve reasonable TempEval-2 scores.

3.5.3 Syntactic and Semantic Information

Syntax is often used alongside lexemes to convey the meaning of an utterance. It is therefore reasonable to investigate the effect of syntactic and semantic information on the temporal relation typing task, as many prior approaches have.

Following [24, 59] add features describing temporal signals, syntactic and semantic roles, and perform reasoning about the context events and timexs appear in to see if they are within one context. They participated in the TempEval challenge, which was not based upon TimeBank but a smaller dataset with a smaller set of potential relation types. They obtain 0.55 accuracy on TempEval's E-E relation typing task using an SVM, which matched the best performance in this task and beat the baseline of 0.47.

During TempEval-1, top performance at event-event relation typing was given by a rule-based system, XRCE-T [60], which relied on deep parsing using a custom parser, XIP. This performance was later matched by a system based on machine learning and notably more complex information sources [61].

Syntactic relations can also play a role in determining temporal relation types. For example, Bethard et al. [62] combine event and syntax features to train an SVM kernel that reaches 89.2 % accuracy on a selected set of event-event relations in TimeBank using a simplified set of three temporal relations. Their feature set includes values that depend upon particular types of syntactic relation between the arguments of a temporal link. Their dataset is constrained to only those event pairs where one event syntactically dominates another.

From TempEval [32], it was observed that performance on tasks that required relation identification between two events or times within the body of the document was low (as opposed to links to the document creation timestamp). One could hypothesise from this that the syntactic structures that connect this pair of lexicalised intervals have some impact on their temporal relation type. To test this hypothesis, [33] created a custom corpus of verb-clause event pairs, using TANGO (see Section A.3.1) and the TimeBank guidelines, with additional annotation rules covering modal/conditional events, aspectual links and permissive verbs (such as 'allow', 'permit' and 'require'). After this, relation identification was modelled using two sets of features; a linguistic set based on event verbs, including things such as tense and aspect and another set based on connecting words (such as signals). This connecting word set included some string features, as well as information about syntactic path and two features based on bags of interconnecting words. Top features were mostly related to target-path (syntactic node path from a clause to its head) or to the subordinated event. Increased word-distance between events decreased relation typing performance, just as was the case in [45].

Cheng et al. [63] use dependency parsing to generate features for relation typing, coupled with a sequence labelling model for events. They assume that, since time is linear, events occur in order, and therefore the events in a document can be treated as a sequence. This leads to an interesting HMM model for inter-event relation typing. Similarly, UzZamana and Allen [64] use a rich, in-depth parser to support their features for a Markov logic network when typing temporal relations. This lead to the best score for event-time labelling in TempEval-2.

As part of a syntacto-semantic approach to temporal information extraction, including timex and event annotation, [65] built on their earlier approach [66] and used syntactic analysis for the event-time relation typing task, also post-correcting classifier output using a system of hand-crafter rules. The approach placed special focus on clause graphs, and achieved moderate success at event-time relation typing.

Ha [67] used a set of lexico-syntactic features for events and times to learn a Markov logic network as a model for temporal relations with a given document. The approach draws additional information from VerbOcean and WordNet. This intuitive approach performs well at event annotation, but extra analysis is required to improve relation typing performance.

Semantic roles have been found to play a useful role in both interval (i.e. event and timex) annotation and temporal relation typing [68]. The concepts are further explored in [69], finding that tense information can be misleading, but still achieving a performance increase over TempEval-2 systems.

3.5.4 Linguistic Context

Some prior approaches rely on discourse information not annotated with TimeML, which typically only applies to a small proportion of tokens in any given text. Looking at the document as a whole, and the linguistic context in which events and timexes lie, may lead to improved relation typing performance (Table 3.7).

VerbOcean is a resource detailing semantic relations between verbs, mined from large corpora. One of these relation types is temporal: *"happens-before"*. Ref. [21]'s system includes experiments which perform VerbOcean (Sect. 3.4.2.1) and GTag[4] rule lookups and use the results as features for machine learning. The data sparsity of VerbOcean leaves it contributing only very slightly to results, to the point where it is hard to tell if performance increases are statistically significant. Out of 24 instances where VerbOcean matches could be made, 19 correctly suggested the final relationship type; 5 incorrect results were found.

The best results are when the scope of TLINKs studied is heavily constrained and situation-specific features used [62, 70]. However, when the features that help in these specific situations are applied generally, they lead to a performance drop in typing of other TLINKs. This suggests that it may be best to apply different typing techniques to particular subsets of TLINKs, instead of trying a "one size fits all" approach.

[4]"GTag takes a document with TimeML tags, along with syntactic information from part-of-speech tagging and chunking from Carafe and then uses 187 syntactic and lexical rules to infer and label TLINKs between tagged events and other tagged events or times." [21].

Of the mechanisms that play a part in conveying temporal relational information, one that has been under-investigated is the use of expressions, typically adverbials or conjunctions, which overtly signal temporal relations – words or phrases such as *after*, *during* and *as soon as*. Very few of the teams participating in the recent TempEval challenges [38–40] exploited these words as features in their automated temporal relation classification systems. Certainly no detailed study of these words and their potential contribution to the task of temporal relation detection has been carried out to date; this is the subject of Chap. 5.

As part of a TempEval system, [71] attempted to find temporal "signal" words – those word which act in a temporal sense to make explicit the nature of a temporal relation, such as *"simultaneously"* – and use these to augment a MaxEnt-based relation labelling system. The approach yielded a mild improvement. Further investigation was given into the impact these signal words can have on the relation typing task [70], showing them to be capable of giving an error reduction of over 50 % for TLINKs that are associated with one. Temporal signals are the focus of a later chapter in this these (Chap. 5).

This has continued through to recent TempEval tasks, such as Clinical TempEval [43, 72], which implements narrative containers as a temporal structuring device [73]. These are defined as "the default interval containing the events being discussed", and implemented in order to increase the informativeness of temporal annotation [74].

Finally, [75] experiment with the addition of event participant and event co-reference features, using an SVM to label relations. This achieves a modest performance level on the event-event relation typing task.

3.5.5 Global Constraint Satisfaction

As temporal relations co-constrain, it can be said that the type of one relation may have a bearing on the types of other relations between which an endpoint is shared. Therefore, considering these global relation type constraints is important to achieving a correct overall relation typing solution, and may lead to improvements in the assignment of individual label types.

Chambers and Jurafsky [51] manually add links to TimeBank v1.2 in cases where events subordinate other events in the same clause (as per [62]) and links between calendar times. They then perform closure and folding over this extended dataset in order to generate extra training examples for an SVM classifier. The output from this classifier is then processed through a model that ensures that temporal relations are globally consistent, correcting relation labels where necessary. No overall accuracy is gained, though after the problem is reduced to just before/after relations, this post-classifier-typing correction yields a 3.6 % accuracy improvement.

Later, [61] use a Markov logic network to model constraints and obtain top accuracy on TempEval's relation typing task. They find that using Markov logic allows better capture of non-absolute rules between relation pairs and that a model need only be built once instead of per-document, which moves focus onto temporal relations instead of the mechanics of machine learning.

3.5.6 *Summary*

Although event-time relation typing accuracies can reach as high as 80 % (as in e.g. TempEval), overall temporal relation typing performance has stalled around 70 % accuracy, leaving temporal relation extraction an open research challenge. Applications require higher performance, but it is not available. Current accuracy is too low to support NLP tasks such as question answering [76], forensic analysis [77] or temporal slot filling [78, 79].

From the above, we can see that classifier choice affects relation typing performance, even for different relation argument types. Including data on global temporal constraints, on syntactic structure and on tense modelling can all help. Further, we see that generic approaches obtain quite different performance in different TLINK settings (such as in TempEval).

Hand-engineering and machine learning methods are effective, even when rule bases have built-in failings. Machine learning methods have reached a performance cap. Improving temporal relation typing accuracy becomes increasingly hard and performance appears to have almost levelled off. Extra effort and sophistication in relation typing approaches yield diminishing returns.

3.5.6.1 TimeML Features

Relying on only the TimeML attribute values as features is not sufficient. Machine learning approaches that use this set of features seem unable to break through the 70 % event-event relation type accuracy barrier, even on folded data [80] or after attempts with a sophisticated array of cutting-edge classifier kernels [81, 82]. Even the introduction of some syntactic information such as argument ancestor path distance and is not sufficient to overcome this barrier [50, 83]. Taking care of other information sources, such as global constraints, yields an immediate but small performance increase over the base feature set [51, 61].

Despite almost a decade of work, relation typing accuracies over even 80 % are a rare event. This is suggestive of some greater difficulty that has not yet been identified. It is possible that there is simply not enough training data, and that generating more through closure is somehow not sufficient (this does not yield performance improvements); this is investigated in Sect. 3.6.1. It could also be the case that TimeML is structurally insufficient somehow, e.g. the markup's attributes and values may be insufficient for capturing all the information required to type a temporal relation. Also, as the highest performance levels are seen on subsets of links from a whole corpus, there may be merit in subsetting relations somehow and working to understand each group. Finally, other problems could arise from the task being insufficiently well-defined, which may manifest in poor inter-annotator agreement. We discuss how well-defined the task is in the rest of this section and relation subsetting in the next chapter.

3.5.6.2 Task Definition Issues

Regarding the definition of the task, there is some data available to describe how well it is understood. In temporal link annotation, separate inter-annotator agreement (IAA) figures are given for relation identification and relation typing. For TimeBank 1.2, relation identification IAA (i.e. the extent to which annotators agreed which pairs of intervals should be related) was low – around 0.55 – though is attributable to the fact that a single temporal relation structure of a document can be described in multiple ways, all equivalent after closure. Unfortunately, IAA figures are not given post-closure, but only pre-closure, and so this 0.55 is a minimum.

Critically, relation type annotation agreement is 0.77 – not absurdly low but below the recommended 0.90 [84]. State-of-the-art in performance overall performance is around 72 % accuracy, which is below IAA, though current performances are nearer to IAA than they are to baseline performance.

There are multiple relationship sets available, and the Allen set used by TimeBank has faced some criticism (e.g. [16]). TempEval-1 and TempEval-2 involved the annotation of data with an alternative (and simpler) relation set. IAA these annotation tasks may be compared to that from TimeBank's to see the impact of reducing the relationship set's complexity on annotator agreement. For TempEval-1, event-time IAA was 0.72 and event-event IAA 0.65. Agreement scores are not readily available for TempEval-2.

When measuring the task difficulty using IAA, it is important to note that not all annotator disagreements are equal. Some relations are temporally equivalent. Disagreeing between SIMULTANEOUS and IDENTITY reduces IAA but the final annotations describe events happening at similar times. Other relations are very close. For example, IBEFORE and BEFORE describe almost the same relationship and temporal ordering. Many relationships place intervals in arrangements where one interval bound is in the same place, but the other is not. When one compares A INCLUDES B with A ENDS B, the start point of interval A is positioned between the start and end points of B – it is only the arrangement of A's end point that these relations disagree upon. TimeML's use of an interval algebra means that the position of both points of both intervals in a relation must be specified. Therefore, it only takes the start or end bound of either of the intervals to be slightly vague for the relationship type to become ambiguous to annotators, fostering annotation disagreement (for details, see the TimeBank corpus notes, e.g. Table A.1).

Table 3.7 Prior work on automatic temporal relation classification. As event-event (E-E) linking is generally a harder task than event-time (E-T) linking, results are in ascending order of event-event relation typing performance. In the case of TempEval results, event-event linking is measured as performance at linking main events in consequent sentences and event-time link is matched to the task of linking events and timexes in the same sentence. Therefore, for TempEval-1, the last two columns correspond to tasks C and A respectively. For TempEval-2, the last two columns correspond to tasks E and C respectively. All TempEval results are for "strict" evaluation

System	Notes	Method	E-E	E-T
Lapata 2006 [85]	BLLIP corpus	Decision tree	70.7	
Gaizauskas 2006 [86]	Clinical corpus	Rule-based	65	
Bramsen 2006 [87]	Medical discharge summaries	Graph based	78.3	
TempEval-1 corpus				
Baseline	*Most common class*		*47*	*57*
Cheng 2007 [63]	Uses dependency parsing	HMM SVM	49	61
Hepple 2007 [47]	Includes text order features	SVM/K*	54	59
Bethard 2007a [88]	Uses syntactic tree features	SVM	54	61
Marsic 2011 [65]		Rule-based		65
Kolya 2011 [55]		CRF + rules		75.9
Puscasu 2011 [66]	Syntactico-semantic features	rule-based	54	**80**
Min 2007 [59]	Focus on rules for marginal cases	SVM	55	58
Kolya 2010 [57]		CRF	55.1	73.8
Hagege 2007 [60]	Based on XIP deep parse data	rule-based	**57**	34
Yoshikawa 2009 [61]	Models global TLINK constraints	MLN	**57**	65
Bethard 2007b [33]	Same-sentence links only	SVM + rules	89.2	
Costa 2013 [89]	Tense, aspect and interval relation rules	Various WEKA	77.9	68.0
TempEval-2 corpus				
Baseline	*Most common class*		*48.63*	*55.07*
Derczynski 2010a [71]	Includes signal information	MaxEnt + rules	45	63
Ha 2010 [67]	Lexico-syntactic feat. + VerbOcean	MLN	51	63
Llorens 2010 [68]	Includes semantic features	CRF	55	55
Kolya 2010 [58]	Includes event head information	CRF	56	63
UzZaman 2010 [64]	Based on TRIPS parse data	MLN	58	65
Hovy 2012 [90]	Tree kernel with bags of [words, PoS tags]	SVM	–	64.5
Laokulrat 2014 [91]	Timegraphs, pairwise entity similarity	Stacked learning	**59.7**	**65.9**

(continued)

Table 3.7 (continued)

System	Notes	Method	E-E	E-T
TimeBank 1.1 corpus				
Baseline	*Most common class*		*33.38*	
Boguraev 2005 [45]	Token windows, FS-grammar features	RRM		53.1
Vasilakopoulos 2005 [52]	Not using folded relations	K*	53.14	
Chambers 2007 [80]	Segregates intra-sent. relations	SVM	**67.57**	
TimeBank 1.2 corpus				
Baseline	*Most common class*		*38.35*	*58.4*
Puscasu 2007 [92]	Maps to TempEval relations	rule-based	53	65
Tatu 2008 [75]	With actor and co-ref features	SVM	58.2	
Mirroshandel 2010 [83]	Bootstrapped kernel w/ AAPD	SVM	66.18	
Chambers 2008 [51]	Models global TLINK constraints	SVM + rules	70.4	
Combined TimeBank 1.2 and AQUAINT TimeML corpus				
Baseline	*Most common class*		*51.57*	*65.3*
Mani 2007 [24]	Uses TimeBank 1.2a	MaxEnt	(59.68)	(82.47)
Mirroshandel 2010a [50]	LICT Polynomial kernel	SVM	67.03	**80.04**
Mirroshandel 2010b [83]	Bootstrapped kernel w/ AAPD	SVM	68.07	
Derczynski 2010b [70]	Signalled TLINKs only	MaxEnt	82.19	

3.6 Analysis

So far, we have shown that general temporal relation typing performance is limited to around the 70 % level (and often not far from the baseline), and that the state of the art isn't moving. This section discusses possible causes, and identifies what does seem to work based on prior efforts.

3.6.1 Data Sparsity

There is not enough annotated data to cover all the combinations of values available through TimeML. This means that there is a chance of seeing new sets of data values that do not exist in any prior labeled dataset. TimeBank has about 6 000 TLINK annotations. Each of these constitutes two arguments (each either a timex or event annotation), a relation type and optionally a reference to text supporting the relation type. Aside from the text that they annotate, events have a class attribute (that has one of seven values), a part-of-speech tag (five choices), a tense (seven choices), an aspect (four choices), and a polarity (two choices) plus cardinality and modality which are free choice (there are 25 values of modality and 15 of cardinality shown in TimeBank). This gives up to 7*5*7*4*2*25*15 = 735 000 possible event configurations (ignoring the free-form lexicalisation of the event). In the simplest case, ignoring event text and text supporting relation types, this makes about $5.4 * 10^{11}$ possible attribute configurations for an event-event temporal relation. The sparseness with which event attribute space and temporal relation attribute are populated by human-annotated corpora means that we are almost certain to encounter previously-unseen combinations of attribute values when attempting the relation typing task on new data. Further, it constrains our ability to make accurate generalisations based on the data that has already been seen.

3.6.2 *Moving Beyond the State of the Art*

To improve performance in the relation typing task, it is important to understand where the problems are and to determine promising directions for further investigation. Some parts of TempEval-1 have been analysed and there are some trends visible even in our small dataset of temporal typing approaches.

Lee [93] provides an error analysis of TempEval-1. Failures are broken down in terms of relation features, such as relation type, argument PoS and tense. It is found that relations of nominalised events are particularly difficult to predict, as are relations where at least one argument is part of reported speech. Data sparsity is a constant problem, with the less-frequent relation types often failing. This error analysis, while enlightening, does not include any attempt to explain or characterise the harder links or to determine if there is a common difficult set.

As for specific tools, Markov logic networks are likely a useful tool for simply modelling global temporal constraints without placing too much restriction or dependency between individual relation labels. They could also help capture knowledge embodied in successful rule-based approaches while being flexible on the known-imperfect rules.

The problem could also lie with representation. The Clinical TempEval series uses narrative containers instead of interval relations; while comparable machine learning performance can be achieved over this representation [94], the inter-annotator agreement issues still stand. It is a hard task [95]. Empirical evaluation suggests that the more expressive representations are harder for statistical learning [96], though insights into human annotation are certainly needed if we are to develop solid temporally-annotated resources.

It is apparent that no single approach has been able to classify a complete set of links; in fact, usually at least a third are mistyped. It would be prudent to conduct an error analysis, in an attempt to characterise the kind of information that one could use to label mislabelled relations. It may be that there is a consistently mislabelled set of "difficult" links within the datasets. Examining these may provide insights in to how to improve temporal relation typing accuracy.

3.7 Chapter Summary

This chapter discussed how we may represent temporal orderings between times and events (temporal intervals). It introduced ideas of point-based, interval-based and semi-interval based temporal relations. A literature review is also included, describing historical and modern systems for automatic annotation of temporal relations. The finding is that general-purpose temporal relation annotation systems have hit a performance ceiling at only modest accuracy. Among other tools, the case is made for a failure analysis of current temporal relation labeling systems.

Descriptions of the concept of a temporal relation, were included offering formal definitions, reasoning algebrae and annotation schemas for temporal relations. These foundations were followed by a review of previous work in automatic temporal relation extraction. It has outlined many sets of approaches, drawing upon statistical methods and rule-based methods; using machine learning and human-engineering systems.

As part of the literature review, evidence was presented that current approaches to the temporal relation typing problem are insufficient and more information than available in the TimeML features may be needed. Further, it is noted that the most successful approaches are those that have focused on a subset of temporal relations that have particular properties. This supports our hypothesis that to understand how to temporally order events described in text, we need to draw upon multiple heterogeneous information sources.

The next chapter will conduct an empirical failure analysis of the link typing task, examining particular subsets of temporal relations and how they may be automatically labelled. Along with a baseline method, these are proposed as avenues of investigation for the later parts of this book.

References

1. Setzer, A., Gaizauskas, R.: A pilot study on annotating temporal relations in text. In: Proceedings of the Workshop on Temporal and Spatial Information Processing, vol. 13, pp. 1–8. Association for Computational Linguistics (2001)
2. Setzer, A.: Temporal information in newswire articles: an annotation scheme and corpus study. Ph.D. thesis, The University of Sheffield (2001)
3. Allen, J.: Maintaining knowledge about temporal intervals. Commun. ACM **26**(11), 832–843 (1983)
4. Hobbs, J.R., Pan, F.: An ontology of time for the semantic web. ACM Trans. Asian Lang. Inf. Process. (TALIP) **3**(1), 66–85 (2004)
5. Prior, A.: Tense logic and the logic of earlier and later. Papers on Time and Tense, pp. 116–134 (1968)

6. Bruce, B.: A model for temporal references and its application in a question answering program. Artif. Intell. **3**(1–3), 1–25 (1972)
7. Moens, M., Steedman, M.: Temporal ontology and temporal reference. Comput. Linguist. **14**(2), 15–28 (1988)
8. Goranko, V., Montanari, A., Sciavicco, G.: A road map of interval temporal logics and duration calculi. J. Appl. Nonclassical Log. **14**(1/2), 9–54 (2004)
9. Denis, P., Muller, P.: Comparison of different algebras for inducing the temporal structure of texts. In: Proceedings of the 23rd International Conference on Computational Linguistics, pp. 250–258. Association for Computational Linguistics (2010)
10. Setzer, A., Gaizauskas, R., Hepple, M.: The role of inference in the temporal annotation and analysis of text. Lang. Res. Eval. **39**(2), 243–265 (2005)
11. McDermott, D.: A temporal logic for reasoning about plans and actions. Cogn. Sci. **6**, 101–155 (1982)
12. Galton, A.: Temporal logic. In: Zalta, E. (ed.) The Stanford Encyclopedia of Philosophy. Stanford University, The Metaphysics Research Lab (2008)
13. Vilain, M., Kautz, H.: Constraint propagation algorithms for temporal reasoning. In: Proceedings of the Fifth National Conference on Artificial Intelligence, pp. 377–382 (1986)
14. Tsang, E.: The consistent labeling problem in temporal reasoning. In: Proceedings of the AAAI Conference, pp. 251–255. AAAI Press (1987)
15. Allen, J.F., Hayes, P.J.: Moments and points in an interval-based temporal logic. Comput. Intell. **5**(3), 225–238 (1989)
16. Freksa, C.: Temporal reasoning based on semi-intervals. Artif. Intell. **54**(1), 199–227 (1992)
17. Kowalski, R., Sergot, M.: A logic-based calculus of events. In: Foundations Of Knowledge Base Management, pp. 23–55. Springer, Heidelberg (1989)
18. Verhagen, M.: Times Between The Lines. Ph.D. thesis, Brandeis University (2004)
19. Derczynski, L., Gaizauskas, R.: Analysing temporally annotated corpora with CAVaT. In: Proceedings of the Language Resources and Evaluation Conference, pp. 398–404 (2010)
20. Denis, P., Muller, P.: Predicting globally-coherent temporal structures from texts via endpoint inference and graph decomposition. In: Proceedings of the 22nd International Joint Conference on Artificial Intelligence (2011)
21. Mani, I., Verhagen, M., Wellner, B., Lee, C., Pustejovsky, J.: Machine learning of temporal relations. In: Proceedings of the 21st International Conference on Computational Linguistics and the 44th Annual Meeting of the Association for Computational Linguistics, p. 760. Association for Computational Linguistics (2006)
22. Pustejovsky, J.: Personal correspondence (2009)
23. Setzer, A., Gaizauskas, R.: Annotating events and temporal information in newswire texts. In: Proceedings of the Second International Conference On Language Resources And Evaluation (LREC-2000), vol. 31, Athens, Greece (2000)
24. Mani, I., Wellner, B., Verhagen, M., Pustejovsky, J.: Three approaches to learning TLINKS in Time ML. Technical Report CS-07-268, Brandeis University, Waltham, MA, USA (2007)
25. Pustejovsky, J., Stubbs, A.: Natural language annotation for machine learning. O'Reilly Media, Inc. (2012)
26. Verhagen, M.: Temporal closure in an annotation environment. Lang. Res. Eval. **39**(2), 211–241 (2005)
27. Chklovski, T., Pantel, P.: Verbocean: Mining the web for fine-grained semantic verb relations. Proc. EMNLP **4**, 33–40 (2004)
28. Chklovski, T., Pantel, P.: Path Analysis for Refining Verb Relations. In: Proceedings of KDD Workshop on Link Analysis and Group Detection (LinkKDD-04) (2004)
29. Lin, D., Pantel, P.: Discovery of inference rules for question-answering. Natural Lang. Eng. **7**(04), 343–360 (2002)
30. Chambers, N., Jurafsky, D.: Unsupervised learning of narrative event chains. In: Proceedings the Annual meeting of the Association for Computational Linguistics. Association for Computational Linguistics (2008)

31. Boguraev, B., Pustejovsky, J., Ando, R., Verhagen, M.: Time Bank Evolution as a Community Resource for TimeML Parsing. Lang. Res. Eval. **41**(1), 91–115 (2007)
32. Verhagen, M., Gaizauskas, R., Schilder, F., Hepple, M., Katz, G., Pustejovsky, J.: SemEval-2007 Task 15: TempEval: temporal relation identification. In: Proceedings of the 4th International Workshop on Semantic Evaluations (SemEval) (2007)
33. Bethard, S., Martin, J., Klingenstein, S.: Timelines from Text: Identification of syntactic temporal relations. In: Proceedings of the International Conference on Semantic Computing, pp. 11–18 (2007)
34. Tannier, X., Müller, P.: Evaluating temporal graphs built from texts via transitive reduction. J. Artif. Intell. Res. **40**, 375–413 (2011)
35. Kelley Jr, J., Walker, M.: Critical-path planning and scheduling. AFIPS Joint Computer Conferences, pp. 160–173 (1959)
36. Levenshtein, V.: Binary codes capable of correcting spurious insertions and deletions of ones. Probl. Inf. Trans. **1**(1), 8–17 (1965)
37. UzZaman, N., Allen, J.: Temporal evaluation. In: Proceedings of the Annual Meeting of the Association for Computational Linguistics: Human Language Technologies, pp. 351–356. Association for Computational Linguistics (2011)
38. Verhagen, M., Gaizauskas, R., Schilder, F., Hepple, M., Moszkowicz, J., Pustejovsky, J.: The TempEval challenge: identifying temporal relations in text. Lang. Res. Eval. **43**(2), 161–179 (2009)
39. Verhagen, M., Saurí, R., Caselli, T., Pustejovsky, J.: SemEval-2010 Task 13: TempEval-2. In: Proceedings of the 5th International Workshop on Semantic Evaluation, pp. 57–62. Association for Computational Linguistics (2010)
40. UzZaman, N., Llorens, H., Derczynski, L., Verhagen, M., Allen, J., Pustejovsky, J.: SemEval-2013 Task 1: TempEval-3: Evaluating Time Expressions, Events, and Temporal Relations. In: Proceedings of the SemEval workshop. Association for Computational Linguistics (2013)
41. Minard, A.L., Agirre, E., Aldabe, I., van Erp, M., Magnini, B., Rigau, G., Speranza, M., Urizar, R.: SemEval-2015 Task 4: TimeLine: Cross-Document Event Ordering. In: Proceedings of the workshop on Semantic Evaluation. Association for Computational Linguistics (2015)
42. Llorens, H., Chambers, N., UzZaman, N., Mostafazadeh, N., Allen, J., Pustejovsky, J.: SemEval-2015 Task 5: QA TempEval. In: Proceedings of the workshop on Semantic Evaluation. Association for Computational Linguistics (2015)
43. Bethard, S., Derczynski, L., Pustejovsky, J., Verhagen, M.: SemEval-2015 Task 6: Clinical TempEval. In: Proceedings of the 9th International Workshop on Semantic Evaluation (SemEval 2015). Association for Computational Linguistics. Association for Computational Linguistics (2015)
44. Popescu, O., Strapparava, C.: SemEval-2015 Task 7: Diachronic Text Evaluation. In: Proceedings of the workshop on Semantic Evaluation. Association for Computational Linguistics (2015)
45. Boguraev, B., Ando, R.: TimeML-compliant text analysis for temporal reasoning. In: Proceedings of International Joint Conference on Artificial Intelligence (IJCAI) (2005)
46. Zhang, T., Damerau, F., Johnson, D.: Text chunking based on a generalization of winnow. J. Mach. Learn. Res. **2**, 615–637 (2002)
47. Hepple, M., Setzer, A., Gaizauskas, R.: USFD: preliminary exploration of features and classifiers for the TempEval-2007 tasks. In: Proceedings of the 4th International Workshop on Semantic Evaluations, SemEval '07, pp. 438–441. Association for Computational Linguistics (2007)
48. Pustejovsky, J., Hanks, P., Sauri, R., See, A., Gaizauskas, R., Setzer, A., Radev, D., Sundheim, B., Day, D., Ferro, L., et al.: The TimeBank Corpus. In: Proceedings of the Corpus Linguistics conference, pp. 647–656 (2003)
49. ARDA: Aquaint timeml corpus (2006). http://www.timeml.org/site/timebank/timebank.html
50. Mirroshandel, S., Ghassem-Sani, G., Khayyamian, M.: Using syntactic-based kernels for classifying temporal relations. J. Comput. Sci. Technol. **26**(1), 68–80 (2010)

51. Chambers, N., Jurafsky, D.: Jointly combining implicit constraints improves temporal ordering. In: Proceedings of EMNLP, pp. 698–706. ACL (2008)
52. Vasilakopoulos, A., Black, W.: Temporally ordering event instances in natural language texts. In: Proceedings of the International Conference on Recent Advances in Natural Language Processing (RANLP 2005) (2005)
53. Mani, I., Schiffman, B., Zhang, J.: Inferring temporal ordering of events in news. In: Proceedings of the Conference of the North American Chapter of the Association for Computational Linguistics on Human Language Technology, pp. 55–57. Association for Computational Linguistics (2003)
54. Reichenbach, H.: The tenses of verbs. Elements of Symbolic Logic. Dover Publications, New York (1947)
55. Kolya, A., Ekbal, A., Bandyopadhyay, S.: Event-time relation identification using machine learning and rules. In: Text, Speech and Dialogue, pp. 117–124. Springer, Heidelberg (2011)
56. Minsky, M.: Logical versus analogical or symbolic versus connectionist or neat versus scruffy. AI Magazine **12**(2), 34 (1991)
57. Kolya, A., Ekbal, A., Bandyopadhyay, S.: Event-event relation identification: a CRF based approach. In: 2010 International Conference on Natural Language Processing and Knowledge Engineering (NLP-KE), pp. 1–8. IEEE (2010)
58. Kolya, A., Ekbal, A., Bandyopadhyay, S.: JU_CSE_TEMP: A first step towards evaluating events, time expressions and temporal relations. In: Proceedings of the 5th International Workshop on Semantic Evaluation, pp. 345–350 (2010)
59. Min, C., Srikanth, M., Fowler, A.: LCC-TE: A hybrid approach to temporal relation identification in news text. In: Proceedings of SemEval-2007, pp. 219–222. ACL (2007)
60. Hagège, C., Tannier, X.: XRCE-T: XIP temporal module for TempEval campaign. In: Proceedings of the fourth international workshop on semantic evaluations (SemEval-2007), pp. 492–495 (2007)
61. Yoshikawa, K., Riedel, S., Asahara, M., Matsumoto, Y.: Jointly identifying temporal relations with Markov logic. In: Proceedings of the Annual Meeting of the Association for Computational Linguistics, pp. 405–413. Association for Computational Linguistics (2009)
62. Bethard, S., Martin, J., Klingenstein, S.: Finding temporal structure in text: machine learning of syntactic temporal relations. Int. J. Semant. Comput. **1**(4), 441 (2007)
63. Cheng, Y., Asahara, M., Matsumoto, Y.: NAIST.Japan: temporal relation identification using dependency parsed tree. In: Proceedings of the 4th International Workshop on Semantic Evaluations, SemEval '07, pp. 245–248. Association for Computational Linguistics (2007)
64. UzZaman, N., Allen, J.: TRIPS and TRIOS system for TempEval-2: Extracting temporal information from text. In: Proceedings of the 5th International Workshop on Semantic Evaluation, pp. 276–283. Association for Computational Linguistics (2010)
65. Marsic, G.: Temporal processing of news: Annotation of temporal expressions, verbal events and temporal relations. Ph.D. thesis, University of Wolverhampton (2011)
66. Puşcaşu, G.: WVALI: Temporal relation identification by syntactico-semantic analysis. In: Proceedings of the 4th International Workshop on SemEval, vol. 2007, pp. 484–487 (2007)
67. Ha, E., Baikadi, A., Licata, C., Lester, J.: NCSU: Modeling temporal relations with markov logic and lexical ontology. In: Proceedings of the 5th International Workshop on Semantic Evaluation, SemEval '10, pp. 341–344. Association for Computational Linguistics (2010)
68. Llorens, H., Saquete, E., Navarro, B.: TIPSem (English and Spanish): Evaluating CRFs and Semantic Roles in TempEval-2. In: Proceedings of the SemEval-2010, pp. 284–291. ACL (2010)
69. Llorens, H., Saquete, E., Navarro-Colorado, B.: Automatic system for identifying and categorizing temporal relations in natural language. Int. J. Intell. Syst. **27**, 680–708 (2012)
70. Derczynski, L., Gaizauskas, R.: Using signals to improve automatic classification of temporal relations. In: Proceedings of the ESSLLI StuS (2010)
71. Derczynski, L., Gaizauskas, R.: USFD2: Annotating temporal expressions and TLINKs for TempEval-2. In: Proceedings of the 5th International Workshop on Semantic Evaluation, pp. 337–340. Association for Computational Linguistics (2010)

72. Bethard, S., Savova, G., Chen, W.T., Derczynski, L., Pustejovsky, J., Verhagen, M.: Semeval-2016 task 12: Clinical tempeval. In: Proceedings of SemEval pp. 1052–1062 (2016)
73. Miller, T.A., Bethard, S., Dligach, D., Pradhan, S., Lin, C., Savova, G.K.: Discovering narrative containers in clinical text. ACL **2013**, 18 (2013)
74. Pustejovsky, J., Stubbs, A.: Increasing informativeness in temporal annotation. In: Proceedings of the 5th Linguistic Annotation Workshop, pp. 152–160. Association for Computational Linguistics (2011)
75. Tatu, M., Srikanth, M.: Experiments with reasoning for temporal relations between events. In: Proceedings of the 22nd International Conference on Computational Linguistics-Volume 1, pp. 857–864. Association for Computational Linguistics (2008)
76. Dang, H., Lin, J., Kelly, D.: Overview of the trec 2006 question answering track. In: Proceedings of the Text Retrieval and Evaluation Conference (2008)
77. Howald, B., Katz, E.: On the explicit and implicit spatiotemporal architecture of narratives of personal experience. Spatial Information Theory, pp. 434–454 (2011)
78. Ji, H., Grishman, R., Dang, H., Li, X., Griffit, K., Ellis, J.: Overview of the TAC2011 Knowledge Base Population Track. In: Proceedings of the Text Analytics Conference (2011)
79. Burman, A., Jayapal, A., Kannan, S., Kavilikatta, M., Alhelbawy, A., Derczynski, L., Gaizauskas, R.: USFD at KBP 2011: Entity linking, slot filling and temporal bounding. In: Proceedings of the Text Analytics Conference (2011)
80. Chambers, N., Wang, S., Jurafsky, D.: Classifying temporal relations between events. In: Proceedings of the 45th Annual Meeting of the ACL on Interactive Poster and Demonstration Sessions, pp. 173–176. Association for Computational Linguistics (2007)
81. Mirroshandel, S., Ghassem-Sani, G.: Temporal relation extraction using expectation maximization. In: Proceedings of RANLP (2011)
82. Mirroshandel, S., Ghassem-Sani, G., Nasr, A.: Active learning strategies for support vector machines, application to temporal relation classification. In: Proceedings of 5th International Joint Conference on Natural Language Processing, pp. 56–64 (2011)
83. Mirroshandel, S., Ghassem-Sani, G.: Temporal relations learning with a bootstrapped cross-document classifier. In: Proceeding of the 19th European Conference on Artificial Intelligence, pp. 829–834 (2010)
84. Hovy, E., Marcus, M., Palmer, M., Ramshaw, L., Weischedel, R.: OntoNotes: the 90% solution. In: Proceedings of the Human Language Technology Conference of the NAACL, Companion Volume: Short Papers, pp. 57–60. Association for Computational Linguistics (2006)
85. Lapata, M., Lascarides, A.: Learning sentence-internal temporal relations. J. Artif. Intell. Res. **27**(1), 85–117 (2006)
86. Gaizauskas, R., Harkema, H., Hepple, M., Setzer, A.: Task-oriented extraction of temporal information: the case of clinical narratives. In: Thirteenth International Symposium on Temporal Representation and Reasoning, TIME 2006, pp. 188–195. IEEE (2006)
87. Bramsen, P., Deshpande, P., Lee, Y., Barzilay, R.: Finding temporal order in discharge summaries. In: AMIA Annual Symposium Proceedings, vol. 2006, p. 81. American Medical Informatics Association (2006)
88. Bethard, S., Martin, J.: CU-TMP: temporal relation classification using syntactic and semantic features. In: Proceedings of the 4th International Workshop on Semantic Evaluations. SemEval '07, pp. 129–132. Association for Computational Linguistics, Stroudsburg, PA, USA (2007)
89. Costa, F., Branco, A.: Temporal relation classification based on temporal reasoning. In: Proceedings of the International Conference on Computational Semantics. Association for Computational Linguistics (2013)
90. Hovy, D., Fan, J., Gliozzo, A., Patwardhan, S., Welty, C.: When did that happen?: linking events and relations to timestamps. In: Proceedings of the 13th Conference of the European Chapter of the Association for Computational Linguistics, pp. 185–193. Association for Computational Linguistics (2012)
91. Laokulrat, N., Miwa, M., Tsuruoka, Y.: Exploiting timegraphs in temporal relation classification. In: TextGraphs-9 (2014)

92. Puşcaşu, G.: Discovering temporal relations with TICTAC. In: Proceedings of the International Conference on Recent Advances in Natural Language Processing (RANLP 2007) (2007)
93. Lee, C., Katz, G.: Error analysis of the TempEval temporal relation identification task. In: SEW-2009 Semantic Evaluations: Recent Achievements and Future Directions, pp. 138–145 (2009)
94. Velupillai, S., Mowery, D.L., Abdelrahman, S., Christensen, L., Chapman, W.W.: BluLab: Temporal information extraction for the2015 clinical TempEval challenge. In: Proceedings of the 9th International Workshop on Semantic Evaluation (SemEval 2015). Association for Computational Linguistics. Association for Computational Linguistics (2015)
95. Scheuermann, A., Motta, E., Mulholland, P., Gangemi, A., Presutti, V.: An empirical perspective on representing time. In: Proceedings of the seventh international conference on Knowledge capture, pp. 89–96. ACM (2013)
96. Derczynski, L.: Representation and Learning of Temporal Relations. Proceedings of the 26th International Conference on Computational Linguistics. Association for Comptuational Linguistics (2016)

Chapter 4
Relation Labelling Analysis

Felix, qui potuit rerum cognoscere causas
Fortunate was he, who was able to determine the causes of things

Georgica (II, v.490)
VIRGIL

4.1 Introduction

In Chap. 3, we discovered that automatic temporal relation typing is a difficult problem. This motivates an investigation into potential ways of improving performance in relation typing. This chapter details an attempt to discover potential ways of improving performance at the task. As humans are readily able to identify the nature of temporal links, one may a priori draw the conclusion that the information required to do so must be available somewhere. This knowledge is in a given document or in information known by the reader before encountering that document (referred to as **world knowledge**). Following the tradition of performing post-hoc analyses on temporally annotated corpora [1, 2], we attempt to characterise and enumerate the in-document knowledge used to support temporal link labelling. In later chapters, we will use some of these types of knowledge to improve automatic temporal relation labelling.

Firstly, this chapter reports on an attempt to identify a common set of challenging temporal links in the TempEval-2 evaluation task. This includes re-examination of the surface information available in TempEval-2 data and an analysis of its distribution in difficult links. Secondly, finding that the surface information presents no clear paths for investigation (as suggested by the performance cap of previous work using surface information discussed in Sect. 3.5.6), a manual investigation of difficult links is undertaken. This comprises a qualitative characterisation of the information used to label the links and motivates our later experimental investigations.

L.R.A. Derczynski, *Automatically Ordering Events and Times in Text*,
Studies in Computational Intelligence 677, DOI 10.1007/978-3-319-47241-6_4

4.2 Survey of Difficult TLINKs

Our hypothesis is that there may be temporal relations that are consistently difficult to classify correctly. That is, some meta-system using an agglomerative approach (e.g. voting) will still have problems with the relation typing problem. It has been difficult to conduct a thorough error analysis of the temporal relation typing task, as authors often do not or cannot make their attempted labellings available, instead publishing more concise overall performance figures. Further, there are many different corpora and corpora-versions used, which hampers comparability.

This section introduces a source of data on attempts at the relation labelling task, followed by a method for grading temporal links in terms of difficulty, reports on the measured proportions of the degrees of difficulty found in typing various temporal relations, defines what constitutes a difficult link and finally presents a data-driven analysis of difficult links based on their surface features.

4.2.1 The TempEval Participant Dataset

As mentioned in Sect. 3.4.4.4, the TempEval exercises strive to produce comparable results over a fixed and agreed dataset, using pre-annotated events, timexes and TLINK arguments, which constrains the scope for variation in systems outside the task focus – temporal labelling methods.

The second TempEval exercise took place in 2010, as part of SemEval [3]. This exercise included four temporal link labelling exercises, in multiple languages, over a purpose-built corpus. Many teams participated in the evaluation and attempted to label these temporal links. As a result, from their submissions we gained a snapshot of the state of the art of temporal link labelling, all on the same data, with multiple approaches. Some teams were prepared to share their submitted results, which, when compared with the correct answer data and the original corpus, could be merged. From this, we were able to measure a "success rate" for each temporal link, determined by the proportion of systems that managed to label it correctly. We then can build a list of links that are difficult for most (or all) of the systems to annotate automatically.

Fortunately, the TempEval-2 organisers released a full dataset of not only source but also evaluation data.[1] Data concerning the distribution of features over events are contained in Figs. 4.1 and 4.2, of features over timexes in Fig. 4.3.

After contacting teams participating in temporal relation labelling tasks, many were kind enough to donate their submitted labels [4–7]. This data was used to conduct a data-driven failure analysis of four separate temporal linking tasks undertaken

[1] Downloadable from http://timeml.org/site/timebank/tempeval/tempeval2-data.zip. It is important to note that this contains more data than was in the tasks set; evaluating systems using this release as-is will not give accurate figures.

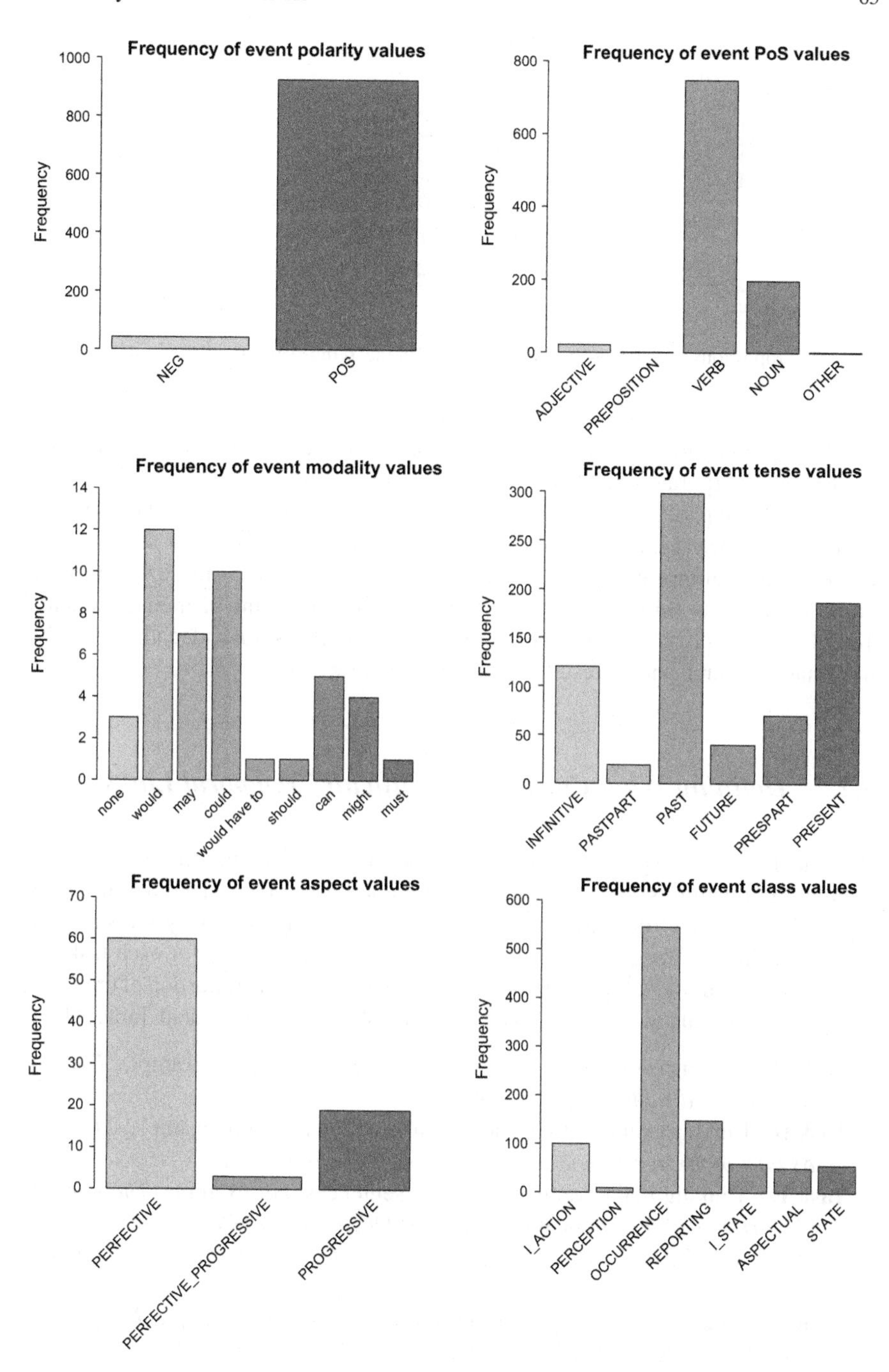

Fig. 4.1 Frequencies of event attribute values in the TempEval-2 English test data

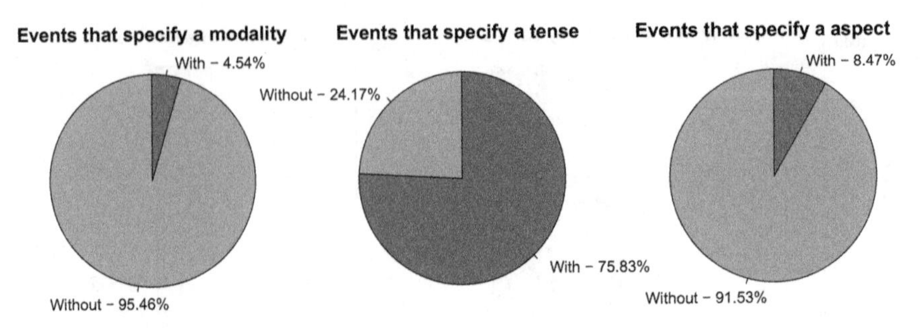

Fig. 4.2 Proportions missing events attribute values in the TempEval-2 English test data

by directly comparable systems. The analysis continues the work on TempEval-1 by [8] and incorporates data from many individual teams.

Given the apparent performance ceiling of systems that use only the annotated TimeML/TempEval-2 feature:value pairs (surface information), clear directions for further investigation are not expected from a formal analysis using these feature:value pairs. However to omit an analysis of difficult links in terms of their arguments' TempEval-2 descriptions would be to ignore a potentially useful and readily available information source and so results are included below.

4.2.2 Defining What Constitutes "difficult" Temporal Links

We start by measuring the "difficulty" of each link, calculating the proportion of attempting labelling systems that generated a correct response. The measurements have values ranging from "all systems correct" (an easy link) to "no systems correct" (a difficult link). This gives a discrete set of difficulty categories for each task. We then count the number of links in each difficulty category as a proportion of the whole and present the data graphically. The results are shown in Fig. 4.4 and Table 4.1.

- **Task C** – Linking events and timexes in the same sentence. For example, *The day$_t$ before Raymond Roth was pulled$_e$ over ...*
- **Task D** – Linking events with the document creation time. For example, *11/01/89$_t$ ... As part of the agreement, Cilcorp said$_e$ it will co-operate.*
- **Task E** – Linking main events in adjacent sentences. For example, *There are 12 flood warnings in the South West, with Met Office warnings for snow covering$_{e1}$ much of the UK. This comes$_{e2}$ just over a week before the start of British Summer Time.*
- **Task F** – Linking main events with subordinate events. For example, *He said$_{e1}$ he discussed$_{e2}$ the issue with Mr. Netanyahu.*

This information permits a brief overall analysis of the relative complexity of the different relation tasks. Task E (Table 4.4) has a fairly stable difficulty gradient,

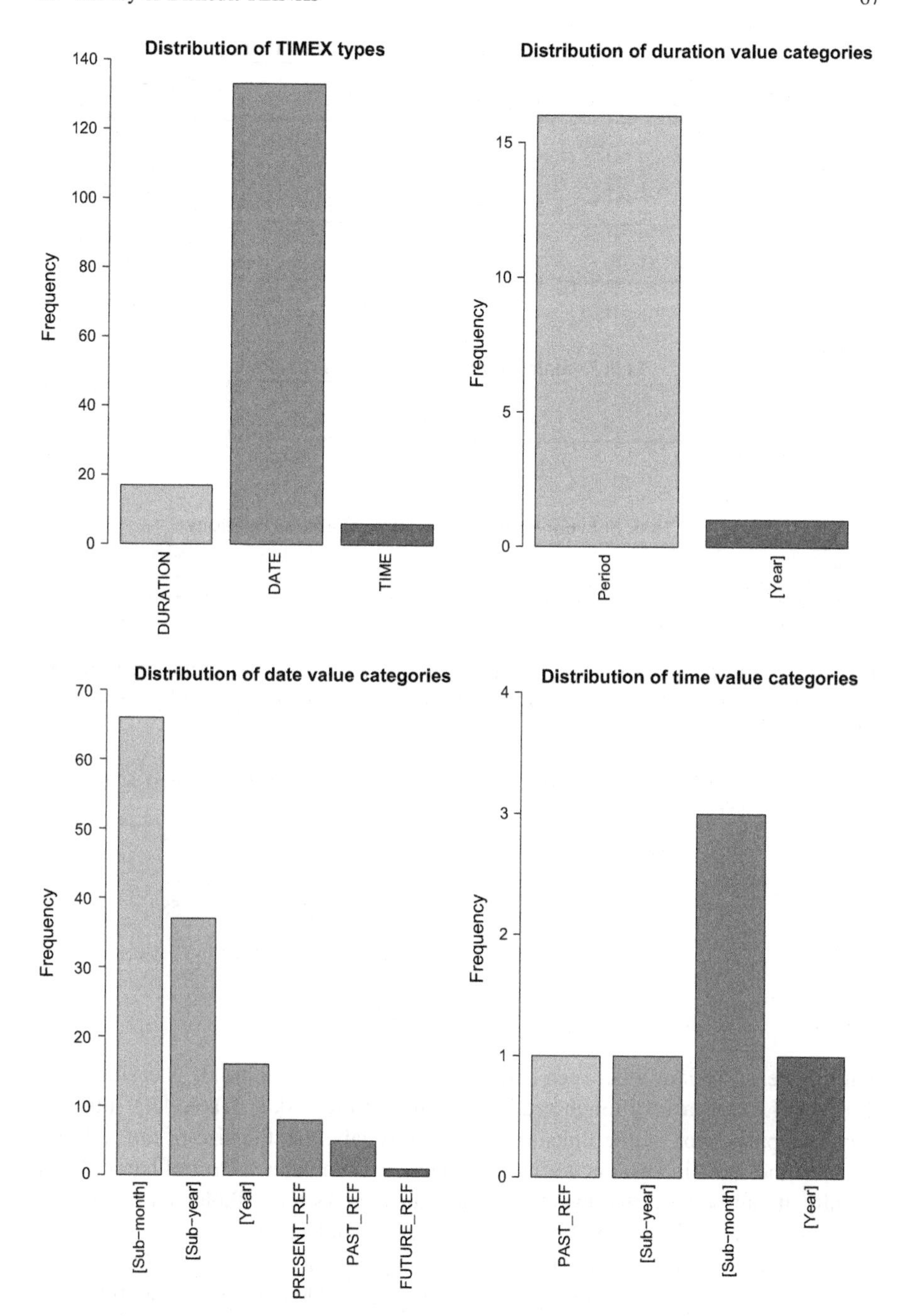

Fig. 4.3 Frequencies of timex attribute values in the TempEval-2 English test data

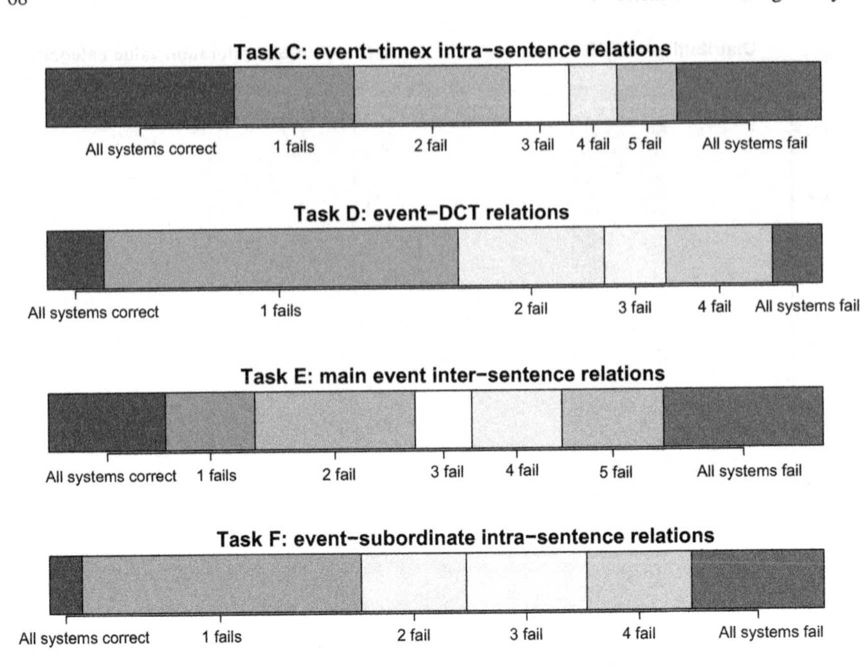

Fig. 4.4 TempEval-2 relation labelling tasks, showing the proportion of relations organised by number of systems that failed to label them correctly. Six systems attempted tasks C and E; five attempted tasks D and F

Table 4.1 Proportion of difficult links in each TempEval-2 task

Task	Difficult links	Difficult proportion (%)	Best score (%)
C	22	8.59	65
D	39	18.4	82
E	62	44.3	58
F	44	46.8	66

with the least deviation between category sizes. Task D (Table 4.3) is easiest. Task C (Table 4.2) has a very tough set; when compared to task E (Table 4.4), although a greater proportion of the links are successfully labelled, the size of the "all fail" group is the same in absolute terms and relatively dominates the set of harder links. Finally, it can be seen that event-event labelling (tasks E+F, Tables 4.4 and 4.5) is harder than event-timex labelling (C+D, Tables 4.2 and 4.3).

Data was available for five or six systems, depending on the task. One system only attempted two of the four tasks, so its absence should not unduly undermine the quality of overall observations. Difficult links are defined as those wrongly labelled by all systems or wrongly labelled by all-but-one system. Given this threshold, we can define a set of difficult links for further analysis. The composition of this set is given in Table 4.1 and shown in Fig. 4.5.

Table 4.2 Error rates in TempEval-2 Task C, event-timex linking

Systems in error	Number of TLINKs	% of TLINKs (%)
No faults	16	24.6
1 fault	10	15.4
2 faults	13	20.0
3 faults	5	7.69
4 faults	4	6.15
5 faults	5	7.69
All fail	12	18.5

Table 4.3 Error rates in TempEval-2 Task D, event-DCT linking

Systems in error	Number of TLINKs	% of TLINKs (%)
No faults	14	7.37
1 fault	87	45.8
2 faults	36	18.9
3 faults	15	15.8
4 faults	26	21.1
All fail	12	6.32

Table 4.4 Error rates in TempEval-2 Task E, linking main events of subsequent sentences

Systems in error	Number of TLINKs	% of TLINKs (%)
No faults	21	15.3
1 fault	16	11.7
2 faults	28	20.4
3 faults	10	7.30
4 faults	16	11.7
5 faults	18	13.1
All fail	28	20.4

Figure 4.6 shows the proportion of links within each task that are difficult and reinforces the earlier observation that event-event links are tougher than event-times links. In the figures, event-timex tasks (C and D) are shown in blue and event-event tasks (E and F) in green. Event-event tasks are comparatively hard, with higher proportions of difficult TLINKs.

Table 4.5 Error rates in TempEval-2 Task F, linking events to events that they subordinate

Systems in error	Number of TLINKs	% of TLINKs (%)
No faults	6	4.26
1 fault	51	36.2
2 faults	19	13.5
3 faults	22	16.1
4 faults	19	13.5
All fail	24	17.5

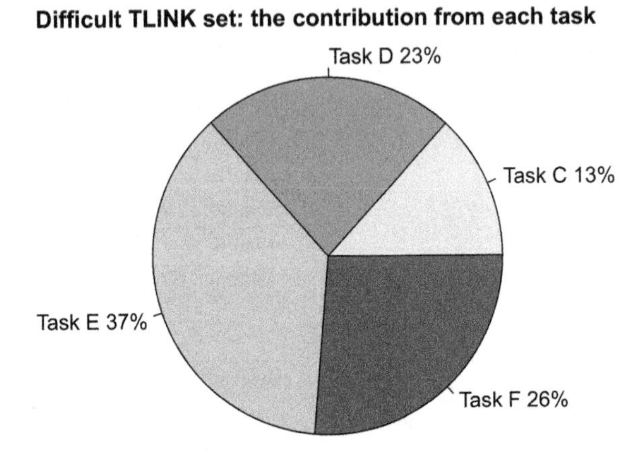

Fig. 4.5 Composition of the set of difficult links. Event-event tasks (E and F) in *green*, event-timex tasks (C and D) in *blue* (color figure online)

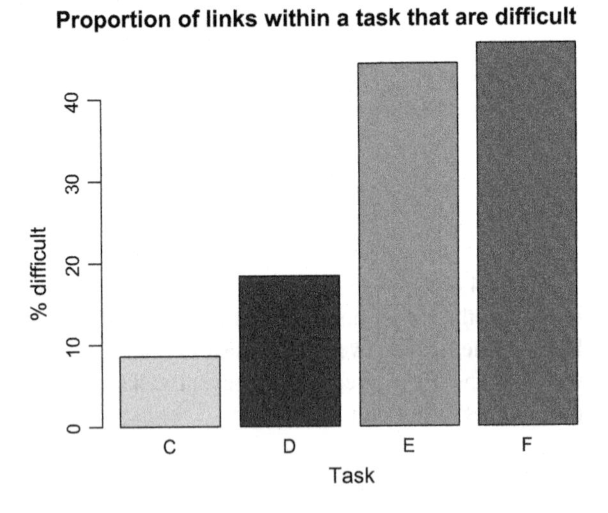

Fig. 4.6 Proportion of each TempEval-2 task's links that are difficult

4.2.3 Comparative Distribution Analysis

Given a set of gold-standard event annotations and gold-standard temporal link annotations, one can conduct a survey of features and values for temporal links. Given also a set of difficult links, one may determine which particular attribute combinations are difficult or easy to automatically label. This is demonstrated in Fig. 4.7, which may be read as follows. Each row corresponds to all events *related to* a given event having a particular property. For example, one row may detail the statistical properties of all other events that are linked to a verb event (e.g. having `pos.VERB`). The columns in this row show the distribution of feature/value pairs in the related event for all relations surveyed. So, continuing the example, in the `pos.VERB` row, the colour represents the likelihood of other argument in the temporal link having a particular feature/value pair. More saturated colours represent higher frequencies. Reds indicate relatively high presence in difficult links (e.g., a "hard" feature combination); blues indicate a low frequency in difficult links (e.g., that the feature combination is "easy").

One could imagine that graph 1 minus graph 2 is graph 3 and that the reds correspond to negative values. Let $\mathbf{A}$ be a matrix of feature:value co-distributions and $\mathbf{B}$ be feature co-distributions in the set of difficult links. If comparison $\mathbf{O} = \mathbf{A} - \mathbf{B}$, then negative values in $\mathbf{O}$ correspond to feature combinations that occur more frequently in $\mathbf{B}$ than $\mathbf{A}$; that is, combinations that are more likely than average to be occur in difficult relations.

4.2.3.1 Difficult Event-Event Link Attribute Distribution

Following this, the Fig. 4.7 presents three saturation maps. The first shows the feature:value co-distribution matrix for all relations. The second shows the matrix just for the difficult relations in that task. By subtracting the second from the first, we can derive the difference between all relations' feature:value distribution and just the difficult relation's distributions. That is, we can identify feature:value pairings that are easier or harder to classify. The harder examples are in red, the easier in blue. Where the distribution varies little between all links and just difficult links, the tone tends to white (unsaturated). Thus, a red cell (for example, where an event of `class.I_STATE` is related to a different event which has `aspect.PERFECTIVE`) represents a frequently difficult combination. Conversely, a dark blue cell (e.g. when an adjective is linked with a present-tense event) shows an easy combination; that is, a pairing which, though frequent, is rarely found in the difficult set. The graphs should not exhibit symmetry, because each row represents a different prior assertion, and is the distribution of other features given that assertion, whereas columns do not represent priors.

This information for Task E, linking main events in successive sentences, is in Fig. 4.7, and for Task F, that of linking events where on linguistically subordinates the other, is presented in Fig. 4.8.

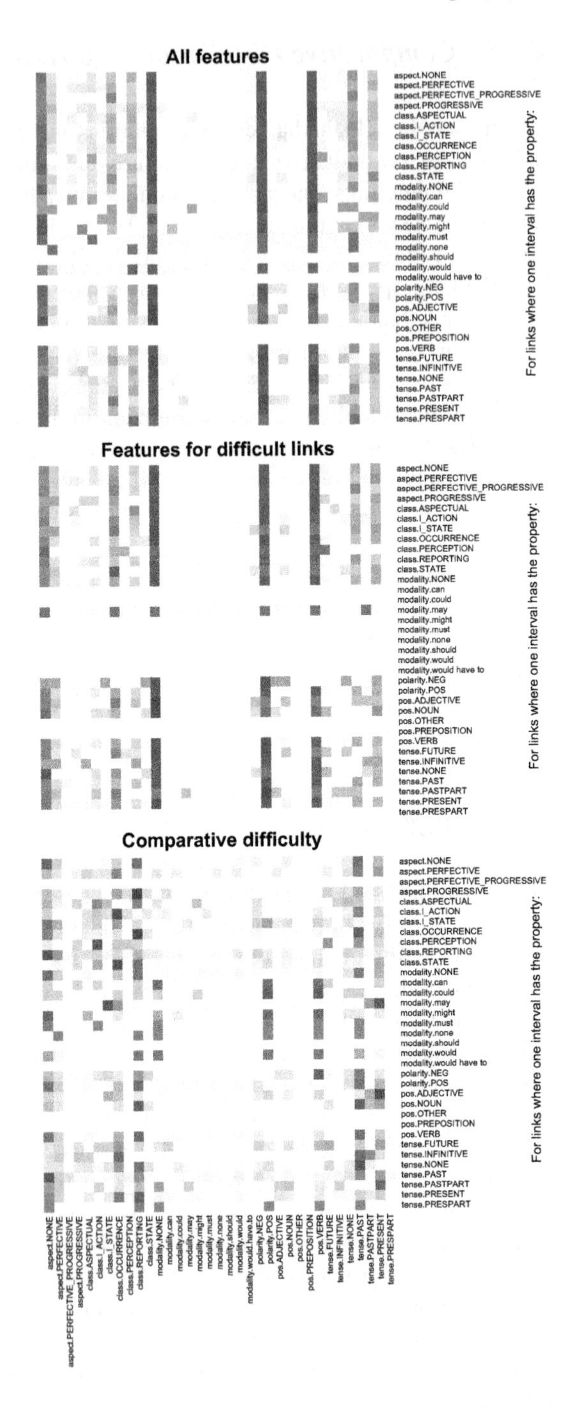

Fig. 4.7 Comparative analysis of features for TempEval-2 task E

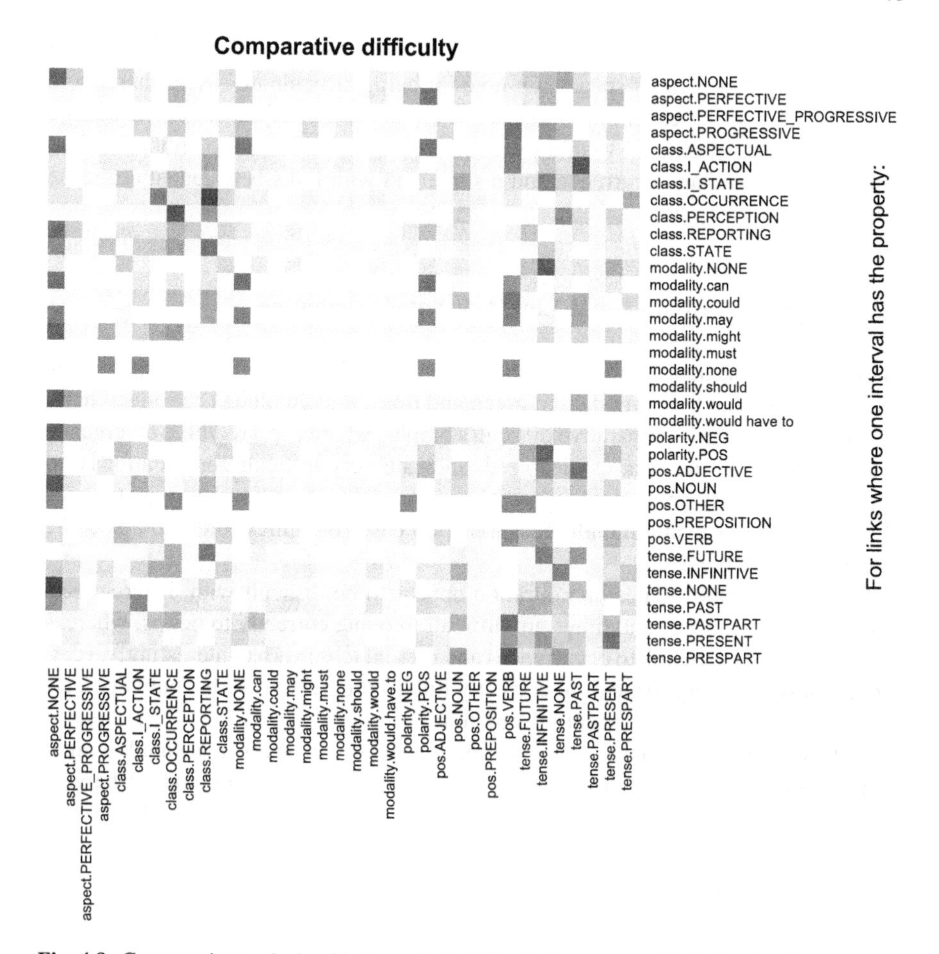

Fig. 4.8 Comparative analysis of features for task F, relating events with their subordinate events

For Task E, from the vertical red stripe in the differential diagram, it can be seen that links to *occurrence*-class events were particularly difficult to label, especially when the other event is of class state or intentional action. However, links to *reporting*-class events were generally easier than average. This could perhaps be due to better consistency in annotations leading to better supervised models, or that a reporting event is typically after the events that are reported but before DCT, giving inherent constraints to this event class. Aside from links with reporting events, particularly easy were links between perceptions and intensional actions (perhaps with perceptions encouraging a reaction?) and links between adjectives and present-tense verbs (perhaps because these always overlap – e.g. *"He says it's hot out there."*).

As for Task F (Fig. 4.8), links with verbs that have no aspect seem to be consistently easier than most. There is less variation in difficulty between certain feature pairings when compared to Task E, as evidenced by the comparatively less saturated graph.

Links to infinitive or un-tensed arguments (e.g. non-verbs) seem to present more difficult than other parts of speech. Of note for being difficult are cases where there is no modality specified in one event and the other is infinitive, possibly due to a reduced number of amodal training examples in a set dedicated to subordination; with links between an occurrence and a state; and with links between future-tense verbs and infinitive verbs.

4.2.3.2 Difficult Event-Timex Link Attribute Distribution

The corresponding data for Tasks C and D are shown in Figs. 4.9 and 4.10 respectively. The colour scheme for event data in green and timex data in blue is continued here, with the exception of comparative difficulty graphs, which use a red/blue divergence colour scheme. In these cases, deep reds indicate very difficult combinations and blue blues very easy ones. Note that the data for task D is only for date-type timexes of granularity less than a month, because in all cases the timex refers to a specific date – DCT – in the data.

For Task C, times, dates and duration appear to be difficult with different sets of event features. Dates and times are difficult to relate correctly to nouns, whereas durations are heard to link to occurrences and present tense verbs. Interestingly, year-sized timexes are very difficult to correctly link to progressive verbs, but very easy to relate to events with no aspect information.

In Task D, we do not have much information. This may be due to a small number of timexes being present in this task's difficult set; the task turned out to be relatively easy. Of these, they are easier to relate correctly to past tensed verbs, and harder to link to occurrence-type events.

4.2.4 Attribute Distribution Summary

It was consistently found that temporal relations between two events are harder to classify than relations between an event and a time. This should direct future research efforts, and was the focus of the latter part of the section, which related a more detailed investigation into the properties of the intervals coupled in difficult links.

Regarding patterns in attribute values over difficult links, although some specific situations of high frequency of difficult links are identified, no clear overall picture emerges. A few specific cases were identified as consistently difficult or easy, but these generally comprised a small proportion of all links. For example, perfect aspect events were had to relate to timexes lasting a year or more; occurrence-class events were difficult to relate with other events, and reporting-class events were easier to relate with other events; and adjective events were easy to relate to present-tense events.

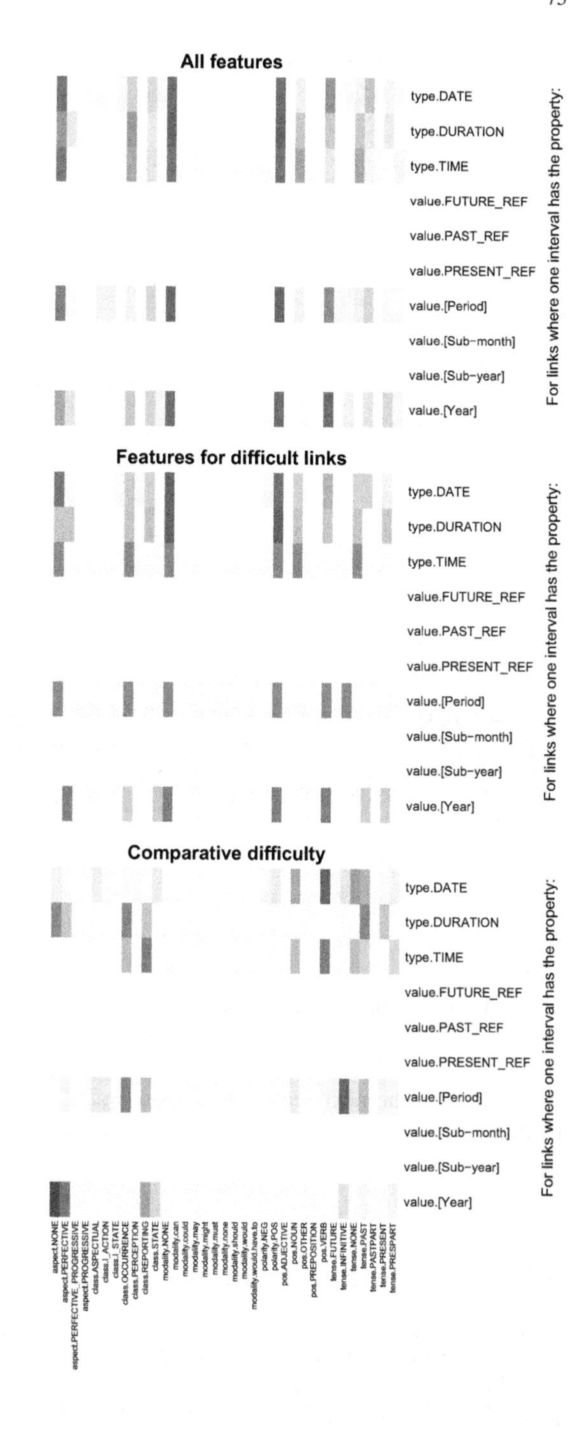

Fig. 4.9 Comparative analysis of features for TempEval-2 task C

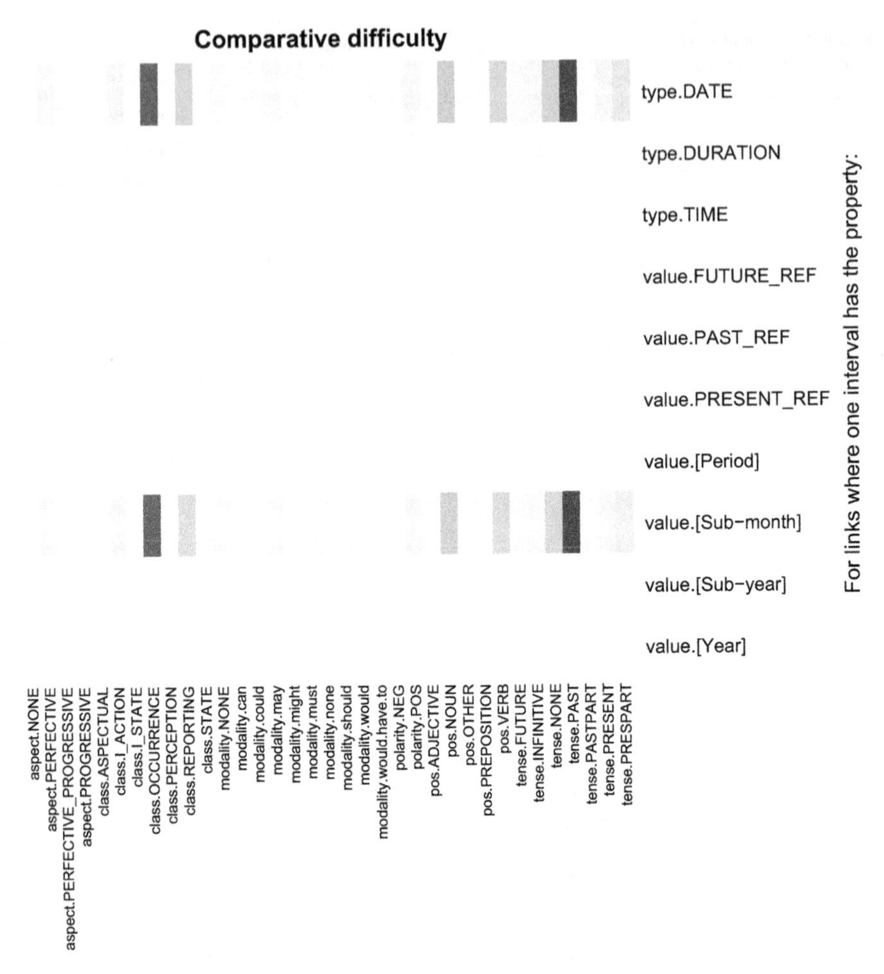

Fig. 4.10 Comparative analysis of features for task D, relating events to DCT

We lead in the next section to a more qualitative approach, taking phenomena contained elsewhere in annotations or not in annotations at all and examining their prevalence in difficult links.

4.3 Extra-Feature Analysis

The overall goal is to determine linguistic sources of temporal ordering information. Because the annotated features do not appear to contain enough information to automatically label links (Sect. 4.2, Chap. 3), other sources of information must be considered. Formal analysis of the surface data does not present immediate clues.

This section presents the results of a survey of each link in the TempEval-2 "difficult set" in terms of the type(s) of information required to determine the temporal relation, aside from that given in TimeML annotations. The resulting information is then used in the next section to attempt to characterise information that temporal links may draw upon, based on prior knowledge about linguistic representations of time.

This analysis was conducted independently of available models and tools, focusing instead on linguistic phenomena. This is to reduce bias from existing methods for and knowledge of the problem. To this end, no TimeML annotation features, tense models or linguistic processing tools were used to construct criteria for characterisation.

4.3.1 Characterisation

It is useful to analyse the difficult TLINKs in a manner that allows identification of common traits. While one can qualitatively express what information is used express a temporal ordering in discourse, to feed into a computational approach one requires quantifiable or at least discrete measures that can be taken consistently from all links. To this end, a set of readily-identifiable linguistic phenomena were determined that could provide temporal information beyond those expressable in TimeML. Each difficult TLINK is then examined and a record made of whether or not each of these phenomena is in place. The result is a survey of types of information used to support temporal orderings for the set of TempEval-2 difficult TLINKs.

The set of phenomena is listed below. Each link may use any number of phenomena. The set is broken into two types: information about the relation and the ordering and information about the interaction between arguments in text.

Relation Information

- **Signalled** - the relation intervals is explicitly expressed by a co-ordinating temporal conjuction or phrase (such as *before*).
- **Inference** - the relation can be easily inferred by reasoning involving other relations in the document
- **From world knowledge** - external information about the general structure of complex events can help determine this relation
- **Iconicity** - temporal order of relation arguments matches the order of their appearance in the source text
- **Disagree** - the annotated relation type is in dispute

Arguments in Text

- **Same sentence** - the relation's arguments are in the same sentence
- **Same clause** - the relation's arguments are in the same clause
- **Tense shift** - there is a shift of tense from one argument to the other
- **Differing modalities** - the arguments do not have the same modality or are not in the same conditional world

- **Differing progression** - one argument is progressive or signifies a culmination or has another aspectual difference from the other
- **Causal** - one argument causes the other and this is critical to the ordering

A "world knowledge" category is therefore included in the above list, in an attempt to roughly estimate how often extra-discourse information is required to resolve difficult links. Also, a "not clear" category is present, for cases where one disagrees with the gold standard.

4.3.2 *Analysis*

The proportion of difficult links that use each of these phenomena as part of their temporal ordering information is shown in Table 4.6.

Overall, 11.2 % of all TLINKs in TimeBank are annotated as using an explicit temporal signal. It seems that a greater-than-average proportion of difficult intra-sentence event-time links rely on signals (task C), but that difficult subordinated relations (task F) use them less often than is typical.

World knowledge rarely supported difficult links. The task that it helped in most was linking main events in adjacent sentences.

Iconicity – that is, when temporal order follows discourse mention order – was generally not observed within the difficult links set. No task had more than 40 % of its difficult links in the same textual and temporal order. The prevalence of iconicity

Table 4.6 Temporal ordering phenomena and their occurrence in difficult links

	Task			
Description	C	D	E	F
Total instances	21	38	62	43
Signalled	33.33 %	13.16 %	11.29 %	6.98 %
Inference	61.90 %	42.11 %	30.65 %	9.30 %
World knowledge	9.52 %	2.63 %	14.52 %	9.30 %
Iconicity	19.05 %	0.00 %	37.10 %	34.88 %
Unclear/Disagree	14.29 %	18.42 %	4.84 %	4.65 %
Same sentence	100.00 %	0.00 %	0.00 %	97.67 %
Same clause	19.05 %	0.00 %	0.00 %	30.23 %
Tense shift	0.00 %	0.00 %	37.10 %	34.88 %
Differing modalities	47.62 %	34.21 %	8.06 %	51.16 %
Differing progression	0.00 %	0.00 %	16.13 %	11.63 %
Causal	0.00 %	0.00 %	9.68 %	4.65 %

was higher in difficult event-event links than event-timex. This may be because it is somewhat redundant in the case of `DATE` and `TIME` timexes, because the timex provides an explicit temporal reference point, and one has less need to rely on implicit factors in order to situate link arguments. Nevertheless, it is interesting to observe that times earlier than events tended to be mentioned in text *after* the events, for the difficult link set. It may also be the case that general discourse follows the principle of iconicity [9] and that having made this observation, automatic temporal relation systems run into difficulties when the principle does not apply.

For event-event links (tasks E and F), a notable proportion of difficult links employ a tense shift. This is where the tense dominating one event is different from that dominating the other. Of the difficult set, this phenomenon occurs 37.1 % of the time in adjacent sentence main event links and 34.9 % of the time in links where one event subordinates another. This suggests that further investigation may be fruitful. There is comparatively very little change of tense in the event-time linking tasks; none in same-sentence event-timex linking and only 5.3 % for event-DCT links.

Differing modalities are very common in in task F's difficult set, as expected for cases where some events subordinate others (this is the category that *if-event-then-event* constructions typically go in), but not common at all for task E.

It is interesting to note the relative lack of shifts in dominant tense in difficult timex-event links when compared to difficult event-event links. This reflects the findings of [10], that temporal adverbs bolster the cognitive role of verb tenses. From these observations, one could suggest that when times are known, a qualifying temporal adverb can be used in place of the information provided by a shift of tense. Validation of this hypothesis remains for future work.

Poor annotation is a potential difficulty source. TempEval-2 data is based on TimeBank, which has an IAA of only 0.77 for TLINK `relTypes`. The TempEval-2 relation set is simpler than TimeBank's, so 0.77 is a minimum IAA. Investigation of the difficult set showed that the frequency of annotation disagreement was in line with what one might expect. The rate of disagreement with the relation type annotation among links in the difficult set was between 4.6 and 18.5 %. This disagreement rate was consistently higher for event-time links than event-event links, but never higher than average IAA accounts for (23 %), so the difficult links are probably not hard due solely to poor annotation.

4.3.3 Signals Versus Tense Shifts

Signals and tense shift are prevalent in the difficult set. It may be useful to investigate both these phenomena. To avoid redundant investigation, one must first establish some degree of independence between the two; if e.g. solving the relation labelling problem for links with tense shifts also solves the problem for those with signals, then it is not worth investigating both.

It has been proposed that both tense shifts and temporal adverbs provide temporal ordering cues [10]. Further, it is suggested that lexicalised temporal markers and tense shifts provide information independently – that is to say, there is no overlap in the information provided by either one of these. Temporal information conveyed by tense shift is independent of that provided in signals. We investigate this using empirical data and briefly test the hypothesis that they are exclusive with regard to the temporal information they provide.

Exploring further the idea of explicit temporal qualification (such as with a temporal adverbial) as an alternative to tense shifts, a brief investigation into the overlap between temporal signals and tense shifts is worthwhile. The data has been gathered and, while not excessive, 105 records (total difficult links from tasks E and F) is enough to estimate the degree of overlap. Results are shown in Table 4.7.

In the case of the difficult event-event links, there was no overlap between links where tense shifted between arguments and links that used an explicit temporal signal. The two categories were in fact mutually exclusive. This was a significant deviation from the overlap that would occur if the two phenomena were mutually exclusive (which would be ~6.3 TLINKs).

Looking at all event-event links in TimeBank 1.2 (difficult and non-difficult), the data is different from TempEval. The overlap between signalled and tense-shifted links is as if these phenomena are almost independent (Table 4.8). This can be demonstrated as follows. The global probability of an event-event link using a signal, $P(S)$, is 7.76 %. Similarly, that of such a link using a tense shift $P(T)$ is 40.6 %. If these variables are independent, $P(S \cap T) = P(S) \cdot P(T)$. We know that in the general case, $P(S \cap T) = 3.30\,\%$; further, $P(S) \cdot P(T) = 3.15\,\%$. This is close to suggesting independence.

Another test is to look for prior probabilities with Bayes' theorem. If independent of T, S with not affect $P(T)$ and vice versa. From the data, $P(T|S) = 42.6\,\%$ which is only 4.9 % out from $P(T)$ and $P(S|T) = 8.11\,\%$ is even closer to $P(S)$ with a 4.5 % difference.

However, for the difficult links, despite $P(S)$ and $P(T)$ having roughly similar values, $P(S \cap T) = 0$, which is significantly different from what one would expect, even after taking into account the size of the dataset. Therefore, we might say that

Table 4.7 Co-occurrence frequencies for temporal signals and tense shifts in event-event difficult links

		Tense shift		
		No	Yes	Total
Signal	No	57	38	95
	Yes	10	0	10
	Total	67	38	105

Table 4.8 Co-occurence frequencies for temporal signals and tense shifts in all TimeBank v1.2's event-event links

		Tense shift		
		No	Yes	Total
Signal	No	1908	1303	3211
	Yes	155	115	270
	Total	2063	1418	3481

having both a tense shift and a signal present makes a link relatively easy to automatically label. Certainly in cases where neither a tense shift not a signal appear, the relation is likely to be difficult to classify.

4.3.4 Extra-Feature Analysis Summary

Certain properties were observed in large proportions of difficult links. Difficult event-time relations (tasks C and D) often employed a temporal signal, relied on global inference, or had differing modalities. Difficult event-event relations (tasks E and F) often relied on inference, exhibited iconicity, involved a tense or aspect shift, or had differing modalities. A large proportion of relations have explicit signal or tense/aspect annotations. As this data is directly available and affects a notable proportion of observed TLINKs, these two phenomena were selected for future investigation.

4.3.5 Next Directions

This section provided a data-driven analysis of difficult TLINKs in a well-known dataset using non-surface criteria. A set of commonly-difficult links was identified for each task. Further, a set of potential temporal information sources was identified in terms of linguistic phenomena and these phenomena monitored for each difficult link. This leads to a set of candidate information types for further investigation. What remains to be done is to outline a framework for working with temporal links using these types of temporal phenomena, so that we have experimental and evaluation methods to use in investigation.

4.4 Analysing TLINKs Through Dataset Segmentation

Our approach is to first identify the type of information used to link two entities and then to classify a relation. This section describes the core approach and then enumerates the various special situations of links to be explored in later experimental chapters.

We are not concerned with determining which entities should be temporally linked in a discourse. We constrain our problem, as in the majority of previous work, to providing the relation type of a given entity pair.

4.4.1 Core Approach

The temporal relation labelling experiments in this book adopt a machine-learning approach, based on that of [11]. Experiments are split into "situations", each of which applies to a subset of temporal links. The identification of links in a particular situation is automatic and a method given for each. Additional features are then added to the core set and a classifier learned and evaluated on the links in a situation. Performance is compared with a classifier learned over the same data but without the additional features.

The base set of features is derived directly from the TimeML attribute values, and is as follows:

- event/timex text;
- TimeML tense for each event;
- TimeML aspect for each event;
- modality for each event;
- cardinality for each event;
- polarity for each event;
- part-of-speech for each event;
- class for each event;
- document function for each timex;
- quantisation for each timex;
- frequency for each timex;
- timex value for each timex;
- temporal function for each timex;
- "mod" for each timex;
- type for each timex;
- are both relation arguments in the same sentence?;
- are both relation arguments in adjacent sentences?;
- if events, do both relation arguments have the same TimeML aspect?;
- if events, do both relation arguments have the same TimeML tense?;
- does argument 1 textually precede argument 2?

4.4.2 Theoretical Assumptions

This analysis expects that expressions conveying temporal relation type are present in discourse. Also, even though each relation may be expressed in many way, we assume that it is not. If every available device above is always used to indicate a temporal relation, the analysis' results would be meaningless, as it would show that all types of information are used for all links.

Instead, the approach outlined above makes the assumption that only the minimum amount of language is used to express temporal information. That is, that information theory [12] concepts such as the minimum description length (MDL) [13] will apply to languages also (as also posited by e.g. [14]). In this context, the MDL principle suggests that unexpected deviations from how time is described require the addition syntactic or lexical information, given a standard "temporal model" of discourse.

Examples of the principle being present in time-relation language are not difficult to come by. One may observe it in phenomena such as temporal signals, tense shifts or temporal expressions. Temporal signals are connectives that explicitly describe a certain ordering but are not required for the majority of relations (they only signal about 12 % of TimeBank's links, for example). Tense shifts require a different term of expression, which may come from the insertion of auxiliary verbs or a change of inflection, and yield a new reference time, event time or even temporal relation. Each shift carries information. Finally, the length and complexity of a temporal expression can correlate to its precision or its distance from the current timeframe; "*At 8.56 am on the 19th August, 2006*" is long, complex and highly specific – "*last week*" serves only to shift the timeframe for anchoring day names backwards. Changing the nominal structure of a sentence is required to express temporal phenomena again. It is this extra information, describing temporal relations, that we are attempting to identify and exploit.

4.5 Chapter Summary

This chapter used a set of empirical data to determine what constitutes a difficult temporal link, and an investigation into linguistic phenomena that occur frequently in the relations that are hardest to automatically label. For each category of relation in TempEval-2 (i.e. Tasks C–F), between 8 and 47 % of temporal relations in documents were difficult for the majority of automatic systems. Event-event relations were consistently the most difficult to type: where 44–47 % of event-event links were difficult, in contrast to event-time links, for which only 8–19 % were difficult.

After an analysis of temporal relations that are difficult to label automatically, themes common in these difficult temporal relations were identified. It was found that two linguistic phenomena were particularly more prevalent in difficult relations than in the general case. First, difficult links often incorporated an explicit co-ordinating temporal signal (a word like *simultaneously* or *thereafter*). Second, shifts of tense and aspect between arguments were often present in difficult links. Other contributing factors were implicit temporal relations discoverable through inference, and changes in modality, though these were less prevalent.

Based on this analysis, the remainder of this book comprises two major parts: an investigation into temporal signals, and another into a framework of tense and aspect. Signals have been found to be useful. We demonstrate how they may be used for temporal relation labelling and then investigate the automatic annotation of temporal signals in Chap. 5. Models of tense can account for a whole group of situations, including reported speech, tense shifts and the use of timexes to shift the frame of reference. Such situations are detailed in Chap. 6.

References

1. Boguraev, B., Pustejovsky, J., Ando, R., Verhagen, M.: TimeBank evolution as a community resource for TimeML parsing. Lang. Resour. Eval. **41**(1), 91–115 (2007)
2. Tissot, H., Roberts, A., Derczynski, L., Gorrell, G., Del Fabro, M.D.: Analysis of temporal expressions annotated in clinical notes. In: Proceedings 11th Joint ACL-ISO Workshop on Interoperable Semantic Annotation (ISA-11), p. 93. Association for Computational Linguistics (2015)
3. Verhagen, M., Saurí, R., Caselli, T., Pustejovsky, J.: SemEval-2010 task 13: TempEval-2. In: Proceedings of the 5th International Workshop on Semantic Evaluation, pp. 57–62. Association for Computational Linguistics (2010)
4. Ha, E., Baikadi, A., Licata, C., Lester, J.: NCSU: modeling temporal relations with markov logic and lexical ontology. In: Proceedings of the 5th International Workshop on Semantic Evaluation, SemEval '10, pp. 341–344. Association for Computational Linguistics (2010)
5. Derczynski, L., Gaizauskas, R.: USFD2: annotating temporal expressions and TLINKs for TempEval-2. In: Proceedings of the 5th International Workshop on Semantic Evaluation, pp. 337–340. Association for Computational Linguistics (2010)
6. UzZaman, N., Allen, J.: TRIPS and TRIOS system for TempEval-2: extracting temporal information from text. In: Proceedings of the 5th International Workshop on Semantic Evaluation, pp. 276–283. Association for Computational Linguistics (2010)
7. Llorens, H., Saquete, E., Navarro, B.: TIPSem (English and Spanish): evaluating CRFs and semantic roles in TempEval-2. In: Proceedings of SemEval-2010, pp. 284–291. ACL (2010)
8. Lee, C., Katz, G.: Error analysis of the TempEval temporal relation identification task. In: SEW-2009 Semantic Evaluations: Recent Achievements and Future Directions, pp. 138–145 (2009)
9. Diessel, H.: Iconicity of sequence: a corpus-based analysis of the positioning of temporal adverbial clauses in English. cognit. linguist. **19**(3), 465–490 (2008)
10. Harris, R., Brewer, W.: Deixis in memory for verb tense. J. Verbal Learn. Verbal Behav. **12**(5), 590–597 (1973)
11. Mani, I., Wellner, B., Verhagen, M., Pustejovsky, J.: Three approaches to learning TLINKS in TimeML. Technical Report. CS-07-268, Brandeis University, Waltham, MA, USA (2007)
12. Shannon, C.: Communication theory of secrecy systems. Bell Syst. Tech. J. **28**(4), 656–715 (1949)
13. Rissanen, J.: Modeling by shortest data description. Automatica **14**(5), 465–471 (1978)
14. Grünwald, P.: A minimum description length approach to grammar inference. Connectionist, Statistical and Symbolic Approaches to Learning for Natural Language Processing, pp. 203–216 (1996)

Chapter 5
Using Temporal Signals

Words are but the signs of ideas.

Preface to the Dictionary
SAMUEL JOHNSON

5.1 Introduction

In Chap. 4, we saw that a proportion of difficult temporal relations were associated with a particular separate word or phrase that described the temporal relation type – a **temporal signal**. The failure analysis in Sect. 4.3.1 finds signals to be of use in over a third of difficult TLINKs. Despite their demonstrable impact on temporal link labelling (see Sect. 3.5.4), no work has been undertaken toward the automatic annotation of temporal signals, and little toward their exploitation. This chapter begins to address these deficiencies.

Temporal signals (also known as temporal conjunctions) are discourse markers that connect a pair of events and times and explicitly state the nature of their temporal relation. Humans resolve events and times in discourses that machines cannot yet automatically label. It is assumed that there must be information in the document and in world knowledge that allows resolution of events, times and relations between them. Temporal signals form part of this information. Intuitively, these words contain temporal ordering information that human readers can access. This chapter investigates the role that temporal signals play in discourse and finds methods for automatically annotating them.

To illustrate:

Example 9 *"The exam papers were submitted before twelve o'clock."*

In Example 9 there is an event, the submitting of exam papers, and a time, twelve o'clock, that are temporally related. The word *before* serves as a signal that describes the nature of the temporal relation between them.

These temporal signals can occur with difficult temporal links and seem to provide explicit information about temporal relation type. It is worth investigating their

L.R.A. Derczynski, *Automatically Ordering Events and Times in Text*,
Studies in Computational Intelligence 677, DOI 10.1007/978-3-319-47241-6_5

potential utility in the relation typing task. If these signals are found to be useful, we may determine how to detect and use them automatically, instead of relying on existing manual annotations. To begin investigation the process of automatic signal annotation, a thorough account of temporal signals is required, followed by an examination of current resources that include temporal signal annotations. Next one may cast the signal annotation problem as a two step process. Firstly, one must know how to determine which words and phrases in a given document are temporal signals. Secondly, one needs to work out with which intervals a given temporal signal is associated, given many candidates. The tasks jointly comprise automatic temporal signal annotation.

This chapter is therefore structured as follows. In Sect. 5.2, we formally introduce background material regarding temporal signals. Section 5.3 reports on the effect that signal information has on an existing relation typing approach compared with the approach's performance sans signal information, finding that adding features that describe temporal signals yields a large error reduction for automatic relation typing. Accordingly, after surveying signal annotations in existing corpora (Sect. 5.4), a method for automatically finding words and phrases that occur as temporal signals is introduced, which first requires the construction of a high-quality ground truth dataset (Sect. 5.5). After developing an approach to finding temporal signal expressions using this new dataset (Sect. 5.6), Sect. 5.7 describes a method for associating temporal signal (once found) with a pair of temporally-related intervals whose relation is described by the temporal signal. The overall performance of the presented temporal signal annotation system is then evaluated. The chapter concludes with an evaluation of the impact this automatic signal annotation has on the overall relation typing task (Sect. 5.8), which is a positive one.

5.2 The Language of Temporal Signals

Signal expressions explicitly indicate the existence and nature of a temporal relation between two events or states or between an event or state and a time point or interval. Hence a temporal signal has two arguments, which are the temporal "entities" that are related. One of these arguments may be deictic instead of directly attached to an event or time; anaphoric temporal references are also permitted. For example, the temporal function and arguments of *after* in "*Nanna slept after a long day at work*" are clear and are available in the immediately surrounding text. With "*After that, he swiftly finished his meal and left*" we must look back to the antecedent of *that* to locate the second argument.

Sometimes a signal will appear to be missing an argument; for example, sentence-initial signals with only one event in the sentence ("*Later, they subsided.*"). These signals relate an event in their sentence with the discourse's current temporal focus – for example, the document creation time, or the previous sentence's main event.

Signal surface forms have a compound structure consisting of a **head** and an optional **qualifier**. The head describes the temporal operation of the signal phrase and the qualifier modifies or clarifies this operation. An example of an unqualified signal expression is *after*, which provides information about the nature of a temporal link, but does not say anything about the absolute or relative magnitude of the temporal separation of its arguments. We can elaborate on this magnitude with phrases which give qualitative information about the relative size of temporal separation between events (such as *very shortly after*), or which give a specific separation between events using a duration as a modifying phrase (e.g. *two weeks after*). In both cases, the signal applies to the ordering of events either side of the separation, rather than the separation itself.

5.2.1 Related Work

Signals help create well-structured discourse. Temporal signals can provide context shifts and orderings [1]. These signal expressions therefore work as discourse segmentation markers [2]. It has been shown that correctly including such explicit markers makes texts easier for human readers to process [3].

Further, words and phrases that comprise signals are sometimes polysemous, occurring in temporal or non-temporal senses. For the purposes of automatic information extraction, this introduces the task of determining when a given candidate signal is used in a temporal sense.

Brée [4] performed a study of temporal conjunctions and prepositions and suggested rules for discriminating temporal from non-temporal uses of signal expressions that fall into these classes. Their approach relies heavily upon the presentation of contrasting examples of each signal word. This research went on to describe the ambiguity of nine temporal prepositions in terms of their roles as temporal signals [5].

Schlüter [6] identifies signal expressions used with the present perfect and compares their frequency in British and US English. This chapter later attempts a full identification of English signal expressions.

Vlach [7] presents a semantic framework that deals with duratives when used as signal qualifiers (see above). Our work differs from the literature in that is it the first to be based on gold standard annotations of temporal semantics and that it encompasses all temporal signal expressions, not just those of a particular grammatical class.

Intuitively, signal expressions contain temporal ordering information that human readers can access easily. Once temporal conjunctions are identified, existing semantic formalisms may be readily applied to discourse semantics. It is however ambiguous which temporal relation any given signal attempts to convey, as investigated by [8] and studied in TimeBank later in this chapter (Sect. 5.4.2). Our work quantifies this ambiguity for a subset of signal expressions.

5.2.2 Signals in TimeML

This section includes work from [9].

TimeML's description of a signal is[1]:

> `SIGNAL` is used to annotate sections of text, typically function words, that indicate how temporal objects are to be related to each other. The material marked by `SIGNAL` constitutes the following:
>
> - indicators of temporal relations such as temporal prepositions (e.g. *"on"*, *"during"*) and other temporal connectives (e.g. *"when"*) and subordinators (e.g. *"if"*). This functionality of the `SIGNAL` tag was introduced by [10].
> - indicators of temporal quantification such as *"twice"*, *"three times"*.

Signals in TimeML are used to mark words that indicate the type of relation between two intervals and also to indicate multiple occurrences of events (temporal quantification). For the task of temporal relation typing, we are only interested in this former use of signals. The annotation guidelines suggest that in TimeML one should annotate a minimal set of tokens – typically just the "head" of the signal.

For example, in the sentence *John smiled after he ate*, the word *after* specifies an event ordering. Example 10 shows this sentence represented in TimeML.

Example 10
```
John <EVENT id="e1"> smiled </EVENT> <SIGNAL id="s1"> after </SIGNAL>
he <EVENT id="e2"> ate </EVENT> .
<TLINK id="l1" eventID="e1" relatedToEvent="e2"
  relType="AFTER" signalID="s1" />
```

TimeML allows us to associate text that suggests an event ordering (a `SIGNAL`) with a particular temporal relation (a `TLINK`). To avoid confusion, it is worthwhile clarifying our use of the term "signal". We use **SIGNAL** in capitals for tags of this name in TimeML and **signal/signal word/signal phrase** for a word or words in discourse that describe the temporal ordering of an event pair. Examples of the signals found in TimeBank are provided in Table 5.1.

It is important to note that not every occurrence of text that could be a signal is used as a temporal signal. Some signal words and phrases are polysemous, having both temporal and non-temporal senses: e.g. *"before"* can indicate a temporal ordering (*"before 7 o'clock"*) or a spatial arrangement (*"kneel before the king"*). This book refers to expressions that could potentially be temporal signals as **candidate signal phrases**. Only candidate signal phrases occurring in a temporal sense are of interest.

The signal text alone does not mean a single temporal interpretation. A temporal signal word such as *after* (for example) is used in TimeBank in `TLINK`s labelled AFTER, BEFORE and INCLUDES. For example, there is no set convention to the order in which a `TLINK`'s arguments should be defined; the AFTER `TLINK` in Example 10 could just as well be encoded as:

```
<TLINK id="l1" eventID="e2" relatedToEvent="e1"
  relType="BEFORE" signalID="s1" />
```

[1] TimeML Annotation Guidelines, http://timeml.org/site/publications/specs.html.

Table 5.1 A sample of phrases most likely to be annotated as a signal when they occur in TimeBank. All corpus data was provided by the CAVaT tool [11]

Phrase	Corpus freq.	Occurrences as signal	Likelihood of being a signal (%)
subsequently	3	3	100
after	72	67	93
's	10	8	80
follows	4	3	75
before	33	23	70
until	36	25	69
during	19	13	68
as soon as	3	2	67

See Table 5.2 for the distribution of relation labels described by a subset of signal words and phrases.

As described above, signals sometimes reference abstract points as their arguments. These abstract points might be a reference time (Sect. 6.3) or an implicit anaphoric reference. As TimeML does not include specific annotation for reference time, one should instead assume that the signal co-ordinates its non-abstract argument with the interval at which reference time was last set. For example, in *"There was an explosion Tuesday. Afterwards, the ship sank"*, we will link the *sank* event with *explosion* (the previous head event) and then associate our signal with this link.

5.3 The Utility of Temporal Signals

Do signals help temporal relation typing? Given the role that they might play in the relation typing task suggested in Sect. 4.3.1 and having a high-level definition of temporal signals, it is next important to establish their potential utility. Since we have in TimeML a signal-annotated corpus, to answer this question, one can compare the performance of automatic relation typing systems with and without signal information. Positive results would motivate investigation into further work on automatic signal annotation. This section relates such a comparison, and includes work from [12]. An extended investigation into this section's findings can be found in [13].

Although accurate event ordering has been the topic of research over the past decade, most work using the temporal signals present in text has been only preliminary. However, as noted in Chap. 3, specifically focusing on temporal signals when

Table 5.2 Signal expressions and the TimeML relations that they can denote. Counts do not match because a single signal expression can support more than one temporal link

Signal expression	TLINK count	AFTER	BEFORE	BEGINS	BEGUN_BY	DURING	ENDED_BY	ENDS	IAFTER	IBEFORE	INCLUDES	IS_INCLUDED	SIMULTANEOUS
after	76	62	3	4				5	2				
when	57	16	3	1	2	1			1	1	9	9	14
until	37	4	7	1			21	1	1	2			
before	36	1	28	2			1	2	1	1			
since	19	9	1	2	7								
already	13		6								4	3	
previously	18	6	12										
while	9												9
meanwhile	9		1			2						1	5
followed	4	2	2										
former	12		12										

Table 5.3 TLINKs and signals in the largest TimeML-annotated corpora

Corpus	Total TLINKs	With SIGNAL	Without SIGNAL
TimeBank v1.2	6418	718 (11.2 %)	5700
AQUAINT TimeML v1.0	5365	178 (3.3 %)	5187
ATC (combined)	11783	896 (7.6 %)	10887
ATC event-event	6234	319 (5.1 %)	5915

classifying temporal relations can yield a performance boost. This section attempts to measure that performance boost.

In TimeML, a signal is either text that indicates the cardinality of a recurring event, or text that explicitly states the nature of a temporal relation. Only the latter sense is interesting for the current work. This class of words and phrases includes temporal conjunctions (e.g. *after*) and temporal adverbials (e.g. *currently*, *subsequently*), as well as set phrases (e.g. *as soon as*). A minority of `TLINK`s in TimeML corpora are annotated with an associated signal (see Table 5.3).

While the processing of temporal signals for `TLINK` classification could potentially be included as part of feature extraction for the relation typing task, temporal signals are complex and useful enough to warrant independent investigation. When the final goal is `TLINK` labelling, once salient features for signal inclusion and representation have been found, one might skip signal annotation entirely and include these features in a temporal relation type classifier. As we are concerned with the characterisation and annotation of signals, we do not address this possibility here, instead attempting to understand signals as an intermediate step towards better overall temporal labelling.

The following experiment explores the question of whether signal information can be successfully exploited for `TLINK` classification by contrasting relation typing with and without signal information. The approach replicated as closely as possible is that of [14], briefly summarised as follows.

The replication had three steps. Firstly, to simplify the problem, the set of possible relation types was reduced (folded) by applying a mapping (see Sect. 3.3.1). For example, as `a` BEFORE `b` and `b` AFTER `a` describe the same ordering between events `a` and `b`, we can flip the argument order in any AFTER relation to convert it to `a` BEFORE relation. This simplifies training data and provides more examples per temporal relation class. Secondly, the following information from each `TLINK` is used as features: event class, aspect, modality, tense, negation, event string for each event, as well as two boolean features indicating whether both events have the same tense or same aspect. Thirdly, we trained and evaluated the predictive accuracy of the maximum entropy classifier from Carafe.[2] To match the original approach, ten-fold cross-validation was used, and a one-third/two-thirds split was also introduced to see the effect of reduced ratio of training:evaluation examples. This split the set of event-event `TLINK`s into a training set of 4156 instances and an evaluation set of 2078 instances.

[2]Available at http://sourceforge.net/projects/carafe/.

Table 5.4 Results from replicating a prior experiment on automatic relation typing of event-event relations

	Corpus	XV accuracy (%)	Train/Eval split (%)	Baseline (%)
Mani et al. results	AQ + TimeBank 1.2a	61.79		51.6
Replicated results	AQ + TimeBank 1.2	60.32	60.04	53.34

In [14], TLINK data came from the union of TimeBank v1.2a and the AQUAINT TimeML corpora. As the TimeBank v1.2a corpus used is not publicly available, we used TimeBank v1.2. This use of a publicly-available version of TimeBank instead of a private custom version was the only change from the previous work. In this work we only examine event-event links, which make up 52.9 % of all TLINKs in our corpus, likely due to minor differences between the TLINK annotations of TimeBank v1.2 and TimeBank v1.2a.

Table 5.4 shows results from replicating the previous experiment on event-event TLINKs. The baseline listed is the most-common-class in the training data. This gives a similar score of 60.32 % accuracy compared to 61.79 % in the previous work. The differences may be attributed to the non-standard corpus that they use. The TLINK distribution over a merger of TimeBank v1.2 and the AQUAINT corpus differs from that listed in the paper.

5.3.1 *Introducing Signals to the Relation Labelling Feature Set*

Now that a reasonable replication of a prior approach has been established, the goal is to measure the difference in relation typing performance that temporal signals make. This requires feature representations of signals. To add information about signals to our training instances, we use the extra features described below; the two arguments of a TLINK are represented by **e1** and **e2**. All features can be readily extracted from the existing TimeML annotations. Only gold-standard signal annotations from the corpora were used.

- **Signal phrase.** This shows the actual text that was marked up as a SIGNAL. From this, we can start to guess temporal orderings based on signal phrases. However, just using the phrase is insufficient. For example, the two sentences *Run before sleeping* and *Before sleeping, run* are temporally equivalent, in that they both specify two events in the order run-sleep, signalled by the same word *before*.
- **Textual order of e1/e2.** It is important to know the textual order of events and their signals even when we know a temporal ordering. Textual order can have a direct effect on the temporal order conveyed by a signal. To illustrate, *"Bob washes before he eats"* describes a story different from *"Before Bob washes he eats"*.

- **Textual order of signal and e1, signal and e2.** These features describe the textual ordering of both TLINK arguments and a related signal. It will also help us see how the arguments of TLINKs that employ a particular signal tend to be textually distributed. The features are required to disambiguate cases where textual order is unreliable. To illustrate, *"Bob washes before he eats"* and *"Before he eats, Bob washes"* describe the same event ordering but have different text orderings.
- **Textual distance between e1/e2.** Sentence and token count between e1 and e2.
- **Textual distance from e1/e2 to SIGNAL.** If we allow a signal to influence the classification of a TLINK, we need to be certain of its association with the link's events. Distances are measured in tokens.
- **TLINK class given SIGNAL phrase.** Most likely TLINK classification in the training data given this signal phrase (or empty if the phrase has not been seen). Referred to as signal **hint**. Referred to as signal **hint**.

5.3.2 *TLINK Typing Results Using Signals*

Table 5.5 shows the results of adding features for temporal signals to the basic TLINK relation typing system. Moving to a feature set which adds SIGNAL information, including signal-event word order/distance data, 61.46 % predictive accuracy is reached. The increase is small when compared to 60.32 % accuracy without this information, but TLINKs that employ a SIGNAL in are a minority in our corpus (possibly due to under-annotation).

The low magnitude of the performance increase seen in Table 5.5 could be due to the way in which training examples are selected. There are in total 11 783 TLINKs in the combined corpus, of which 7.6 % are annotated including a SIGNAL; for just TimeBank v1.2, the figure is higher at 11.2 % (see Table 5.3 and also Fig. 5.1). The proportion of signalled TLINKs in our data – event-event links in the combined AQUAINT/TimeBank 1.2 corpus – is lowest at 5.1 %. It is possible that signalled TLINKs are classified significantly better using this extended feature set, but account for such a small part of this dataset that the overall difference is small. To test this, the experiment is repeated, this time splitting the dataset into signalled and non-signalled TLINKs.

Table 5.5 TLINK classification with and without signal features, using both 10-fold cross validation and a one-third/two-thirds split between evaluation and training data

Predictive accuracy	XV	Split (%)
Baseline (most common class)	53.34 %	53.34
Without signal features	60.32 %	60.04
With basic signal features	**61.46 %**	60.81
With signal features including hint	n/a	**61.98**

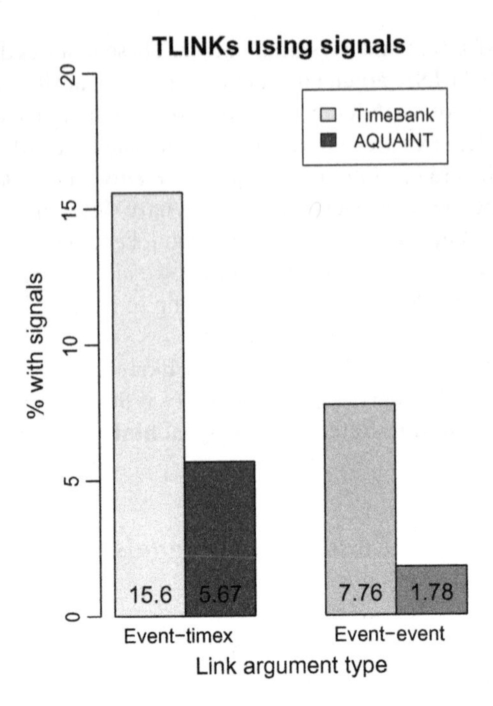

Fig. 5.1 Signalled TLINKs by argument type (event-event or event-tlink) in TimeBank 1.2 and the AQUAINT TimeML corpus. The *paler columns* correspond to TimeBank, the *darker* AQUAINT

If there is no performance difference between feature sets when classifying TLINKs that *do* use signals, then our hypothesis is incorrect, or the features used are insufficiently representative. If signals are helpful, and our features capture information useful for temporal ordering, we expect a performance increase when automatically classifying signalled TLINKs. Results in Table 5.6 support our hypothesis that

Table 5.6 Predictive accuracy from Carafe's maximum entropy classifier, using features that do or do not include signal information, over signalled and non-signalled event-event TLINKs in ATC. The baseline is accuracy when the most-common-class is always assigned

	Cross validation		Train/Eval split	
Predictive accuracy	Unsignalled (%)	Signalled (%)	Unsignalled (%)	Signalled (%)
Baseline (most common class)	52.68	62.41	52.68	62.41
Plain features	**62.05**	55.65	61.81	60.32
Plain, signal features	62.05	**69.57**	61.81	**82.19**
Plain, signal features, hint	62.05	41.72	–	–

signals are useful, but we are performing nowhere near the maximum level suggested above. Data sparsity is a problem here, as the combined corpus only contains 319 suitable TLINKs, and both source corpora show evidence of signal under-annotation. The results also suggest that the signal hint feature was not helpful; this is the same result found by [15].

Exploring the strongest feature set (basic+signals; no hint), and attempting to combat the data sparsity problem, we used 10-fold cross validation instead of a split; results are also in Table 5.6. This again shows a distinct improvement in the predictive accuracy of signalled TLINKs using this feature set over the features in previous work. Cross-validation also gives better overall accuracy. This is likely because of the low volumes of training data mean that the real difference in number of examples between 10-fold cross validation and a one-third/two-thirds split can make a large contribution to classifier performance.

5.3.3 Utility Assessment Summary

When learning to classify signalled TLINKs, there is a significant increase in predictive accuracy when features describing signals are introduced. This suggests that signals are useful when it comes to providing information for classifying temporal links, and also that the features we have used to describe them are effective.

Now that it is confirmed that signals are helpful in temporal relation typing, the next task is to determine how to annotate them automatically. A good account of existing resources may give clues for this process. After this, one needs to explore how to discriminate whether or not a candidate signal expression is used as a temporal signal in text. Next, after finding a temporal signal, we need to determine which intervals it temporally connects. Finally, we can attempt to annotate a temporal link based on the signal.

5.4 Corpus Analysis

In order to understand temporal signals, this section investigates the role of hand-annotated temporal signals in the TimeBank dataset. Further, casual examination reveals that words acting in a temporal signal role in existing datasets are not always annotated as such. Under-annotation can depend on how well the annotator understands the task, and the clarity of annotation guidelines. This section discusses the TimeML definition of signals and describes an augmented corpus which has received extra annotation.

Using the TimeBank corpus, we set out to answer the following questions:

1. Of the expressions which can function as temporal signals, what proportion of their usage in the TimeBank corpus is as a temporal signal? E.g. how ambiguous are these expressions in terms of their role as temporal signals?
2. Of the occurrences of these expressions as temporal signals, how ambiguous are they with respect to the temporal relation they convey?

The following section (which includes material from [9]) provides provisional answers to these questions – provisional as one of the difficulties we encountered was significant under-annotation of temporal signals in TimeBank. We have addressed this to some extent, but more work remains to be done. Nonetheless we believe the current study provides important insights into the behaviour of temporal signals and how they may be exploited by computational systems carrying out the temporal relation detection task.

5.4.1 *Signals in TimeBank*

The TimeML `<SIGNAL>` element bounds a lexicalised temporal signal. Summary information on the `SIGNAL` elements in TimeBank 1.2 is in Table 5.7 and the number of links per signal in Table 5.8. Although permitted under TimeML 1.2.1 for denoting cardinality, no signals have been assigned to event instances for this purpose, although there is one unassigned signal annotation that does indicate event cardinality.

Table 5.7 How `<SIGNAL>` elements are used in TimeBank

Annotated `SIGNAL` elements	758
Signals used by a `TLINK`	721
Signals used by an `ALINK`	1
Signals used by a `SLINK`	39
`TLINK`s that use a `SIGNAL`	787
Signals used by more than one `TLINK`	54

Table 5.8 The number of TLINKs associated with each temporal signal word/phrase, in TimeBank. Signals not used on TLINKs (e.g. those used on aspectual or subordinate links, or for event cardinality) are excluded. The distribution appears to be Zipfian [16]

Argument pairs co-ordinated	Frequency
1	597
2	41
3	12
5	1

In cases where a specific duration occurs as part of a complex qualifier-head temporal signal, e.g. *two weeks after*, TimeBank has followed the convention that the signal head alone is annotated as a SIGNAL and the qualifier is annotated as a TIMEX3 of type DURATION.

5.4.2 Relation Type Ambiguity

The nature of the temporal relation described by a signal is not constant for the same signal phrase, though each signal tends to describe a particular relation type more often than other types. Table 5.2 gives an excerpt of data showing which temporal relations are made explicit by each signal expression. The variation in relation type associated with a signal is not as great as it might appear as the assignment of temporal relation type has an element of arbitrariness – one may choose to annotate a BEFORE or AFTER relation for the same event pair by simply reversing the temporal link's argument order, for example. There is no TimeML convention regarding how TLINK annotation arguments should be ordered. Nevertheless, it is possible to draw useful information from the table; for example, one can see that *meanwhile* is much more likely to suggest some sort of temporal overlap between events than an ordering where arguments occur discretely.

5.4.2.1 Closed Class of Signals

To what extent are the words sometimes annotated as temporal signals in TimeBank actually used as time relaters?

As temporal signals and phrases are likely to be a closed class of words, our approach is to first define a set of temporal signal candidate words. For each occurrence of one of these words in a discourse, we will decide if it is a temporal signal or not.

Because they do not contribute to temporal ordering, annotated signals that indicate the cardinality of recurring events were removed before experimentation. We have derived a closed class of 102 signal words and phrases from [17] (see for example Sect. 10.5, "Time Relaters"), given in Table 5.9. This list is long but may not be comprehensive. Automatic signal annotation can be approached by finding words in a given document that are both within this closed class of candidate signal phrases and also occur having a temporal sense. TimeBank contains 62 unique signal words and phrases (ignoring case), annotated in 688 SIGNAL elements and used by 718 TLINKs. Of these 62, over half (39) are also found in our list above. The remaining 23 signals correspond to only 45 signal mentions, supporting 46 temporal links. Thus, if we can perfectly annotate every signal we find in text based on our closed class, we will have described 93.1 % of TLINK-supporting signals and be better able to label 93.6 % of TLINKs that have a supporting signal.

Table 5.9 A closed class of temporal signal expressions

after	ensuing	meantime	soon
afterwards	eventually	momentarily	still
again	fifthly	next	subsequent
already	finally	ninethly	subsequently
as	first	now	succeeding
as soon as	firstly	nowadays	suddenly
as yet	following	on	supervening
at	for	once	then
at once	forever	originally	thereafter
at this point	for ever	over	thirdly
before	former	past	through
beforehand	formerly	preceding	throughout
between	fourthly	presently	til
by	frequently	previous	till
coexisting	from	previously	to
coinciding	here	prior	up to
concurrent	hitherto	recently	until
concurrently	immediately	secondly	when
contemporaneous	in	seventhly	whenever
contemporaneously	initially	shortly	while
contemporary	instantly	simultaneous	whilst
directly	last	simultaneously	within
during	late	since	yet
earlier	lately	sixthly	's
early	later	so long as	
eighthly	meanwhile	sometime	

To provide a surface characterisation of the role signals play, the distribution of their part of speech tag (from PTB) over signals in TimeBank is given in Table 5.10. Many uses are as prepositions, perhaps for attaching events to each other by means of prepositional phrases.

Of the closed class entries detailed in Table 5.9, 25 entries occur in the corpus but are never annotated as signal text: *again*, *directly*, *early*, *finally*, *first*, *here*, *last*, *late*, *next*, *now*, *recently*, *eventually*, *forever*, *formerly*, *frequently*, *initially*, *instantly*, *meantime*, *originally*, *prior*, *shortly*, *sometime*, *subsequent*, *subsequently* and *suddenly*.

We could also derive an alternative signal list by extracting all phrases that are found as the first child of SBAR-TMP constituent tags, as suggested in Dorr and Gaasterlaand [18]. For example, in Fig. 5.2 (an automatically parsed and function-tagged sentence from TimeBank's wsj_0520.tml), the first child of the SBAR-TMP constituent is a one-leaf IN tag. The text is *after*, which we would treat as

Table 5.10 Distribution of part-of-speech in signals and the first word of signal phrases

Part of speech	Frequency	Proportion (%)
IN	521	77.3
RB	73	10.8
WRB	53	7.9
JJ	14	2.1
RBR	5	0.7
VBG	4	0.6
CC	2	0.3
RP	1	0.1
JJR	1	0.1

Fig. 5.2 An example SBAR-TMP construction around a temporal signal

```
(S1 (S (NP-SBJ (NNP Nashua))
      (VP (VBD announced)
        (NP (DT the) (NNP Reiss) (NN request))
        (SBAR-TMP (IN after)
          (S (NP-SBJ (DT the) (NN market))
            (VP (VBD closed))))) (. .)))
```

Table 5.11 The set of signal words and phrases suggested by the SBAR-TMP model, broken into correctly and incorrectly detected phrases

Correct examples	Incorrect examples
after	at least
as	as surely
before	several months
once	nearly two months
since	even
until	only
while	soon
when	

a temporal signal. This approach returns a restrictive set of temporal signals, shown in Table 5.11, though contains few false positives.

5.4.3 Temporal Versus Non-temporal Uses

The semantic function that a temporal signal expression performs is that of relating two temporal entities. However, the words that can function as temporal signals also play other roles.

For example, one may use *before* to indicate that one event happened temporally prior to another. This word does not always have this meaning.

Example 12 "I will drag you before the court!"

In Example 12, the reading is that one will be summoned to appear in front of the court – the spatial sense – and not that the reader will be dragged, and then later the court will be dragged. It is important to know the correct sense of these connective words and phrases.

Of all temporal relations (`TLINK`s) in the English TimeBank, 11.2 % use a temporal signal in the original annotation (Table 5.3). It is important to note that some instances of signal expressions are used by more than one temporal link; see Table 5.8 for details. The most frequent signal word was "in", accounting for 24.8 % of all signal-using `TLINK`s. However, only 13.3 % of occurrences of the word "in" have a temporal sense. The word "after" is far more likely to occur in a temporal sense (91.7 % of all occurrences).

As an aside, the notion that temporal signals might be easily picked out based upon word class may be dispelled by examining the distribution of parts-of-speech possessed by temporal signals – see Table 5.10. Part of speech is not a reliable disambiguator of sense, in this case.

5.4.4 Parallels to Spatial Representations in Natural Language

Time and space are related and often an event will be positioned in both. Language used for describing time and language used for describing space are often similar, not least in the fact they they both use signals and often even use the same words as signals. Temporal signals relate a pair of temporal intervals, and spatial signals relate a pair of regions. Although not the focus of this chapter, it is useful to note the common and contrasting behaviours of temporal and spatial signals that emerged during investigation.

SpatialML [19] is an annotation scheme for spatial entities and relations in discourse.[3] Among other things it includes elements for annotating relations between spatial entities.

Links in SpatialML may be topological or relative. Topological links include containment, connection and other links from a fixed set based on the RCC8 calculus. SpatialML relative links, on the other hand, express spatial trajectories between locations.

In the revised ACE 2005 SpatialML annotations,[4] 97.5 % of all `RLINK`s (the SpatialML representation for a relative spatial link) have at least one accompanying textual signal (See Table 5.12). Compared to TimeBank's 11.2 % of TLINKs having a signal, SpatialML relative links are much more likely to use an explicit signal

[3] Although SpatialML has now been superseded by ISO-Space, we are concerned in this section with a SpatialML annotated corpus; there is no ISO-Space equivalent at the time of writing.

[4] LDC catalogue number LDC2011T02.

Table 5.12 Frequency of signal usage for different types of spatial link in the ACE 2005 English SpatialML Annotations Version 2

Link type	SpatialML element	Occurrences	Signalled	Signalling rate (%)
Relative	RLINK	80	78	97.5
Topological	LINK	378	7	1.85

than TimeML temporal relations. This may be because the mechanisms available in language for expressing temporal relations are wider than those for relating spatial entities. For example, to relate events in English, one may choose to use a tense and aspect (which involves inflection or added auxiliaries) instead of adding a signal word. Furthermore, there are three spatial dimensions in which to describe an entity; in contrast, the arrow of time supplied a single unidirectional dimension, which limits range of movements and relations available.

Unlike with relative links, signal usage is lower with topological links. Only 1.85 % of the latter use a signal. This distinction between relative and the temporal equivalent of topological links is not made in TimeML.

This difference in signal usage rate between topological and relative links may be because topological links are used to express relations that we infer from world knowledge and do not lexicalise. In *"A Ugandan village"*, one does not need to explain that the village is in Uganda. Relative links define one region relative to another. The nature of the relation is not easy to discern and so needs to be made explicit.

Because of the dominance of spatio-temporal sense frequencies over other uses of many of the words in this class, work on temporal signals may provide insights for future researchers working on determining spatial labels using spatial signals. This chapter will later (Sect. 5.6.4.3) on show how indications of spatial signal usage help discern temporal from non-temporal candidate signal words.

5.5 Adding Missing Signal Annotations

Given an idea of what signals are and evidence of their utility in temporal relation typing, the next step was to attempt automatic signal annotation. This was a two stage process, first concerned with identifying signal expressions that occur in a temporal sense, and then with determining which pair of events/timexes any given temporal signal co-ordinates. A preliminary approach to finding temporal signal expressions found that the dataset used suffered from low annotation quality, and so after outlining the preliminary approach, this section focuses on how the resources could be (and were) improved.

Upon examination of the non-annotated instances of words that usually occur as a temporal signal (such as *after*) it became evident that TimeBank's signals are under-annotated. In an effort to boost performance, and as there is evidence of annotation errors in the source data, we revisited the original annotations.

This chapter outlines the signal expression discrimination task only briefly, instead focusing on corpus re-annotation. The next section is dedicated entirely to the discrimination problem.

5.5.1 Preliminary Signal Discrimination

The overall problem is to find expressions in documents that occur as temporal signals (a fuller problem definition is given below, in Sect. 5.6). This was approached by considering all occurrences of expressions from the above closed class of expressions (e.g. candidate signals) and judging, for each instance, whether or not it had a temporal sense. Judgement was performed by a supervised classifier (maximum entropy), trained and evaluated using cross-validation, based on the features listed in Sect. 5.6.4.2.

Failure analysis of this initial approach suggested that the corpus was too poorly annotated to serve either as representative, solid training data for signal discrimination, or for an evaluation set for a signal discrimination approach. Some re-annotation was necessary to improve the quality of the ground truth data. This section relates the approach to, and results of, that re-annotation.

5.5.2 Clarifying Signal Annotation Guidelines

Given that the signal annotations in TimeBank are not of sufficient quality, there are three potential causes for this: annotator fatigue, insufficient annotation guidelines, or a poor definition of signals. As annotator fatigue depends on the method of an individual annotation exercise, and TimeML's signal definition is sufficient, we seek to clarify the annotation guidelines.

To clarify the guidelines, it's important to have a thorough definition of temporal signals. While TimeML's definition is sufficient, this chapter offers an extended definition of temporal signals in Sect. 5.2.

Signal surface forms have a compound structure of a **head** and an optional **qualifier**. The head describes the general action of the signal phrase and may optionally have an attached modifying phrase. Only the head should be annotated.

Example 13 "I arrived long after the party had finished."

In Example 13, the word *after* is annotated, and the qualifier *long* is not. This would be annotated in TimeML something like:

```
I arrived long <SIGNAL>after</SIGNAL> the party had
finished.
```

Further, a temporal signal has two arguments, which are timexes or events which are temporally related. Often both of these are explicit in the text immediately surrounding the signal. However, one may be elsewhere, as an implied argument.

5.5.3 *Curation Procedure*

The goal is to create a firm ground truth for further investigation. Given the extended definition of a signal and the guideline clarifications just mentioned, this section details the ensuing exercise of hand-curating TimeBank to repair signal annotations.

A subset of signal words was selected for re-annotation. All instances of these words (both as temporal and non-temporal) were re-annotated with TimeML, adding `EVENT`s, `TIMEX3`s and `SIGNAL`s where necessary to create a signalled `TLINK`. We will reference this version of TimeBank with curated signal annotations as **TB-sig**.

Evaluating correct classifications against erroneous reference data will lead to artificially decreased performance. To verify that the training data (which is also evaluation data for cross-validation) is from a correct annotation, negative examples of signal words were checked manually. False negatives are removed by annotating them as TimeML signals, associating them with the appropriate TLINK or adding TLINKs and EVENTs where necessary.

Checking the entire corpus would be an exhaustive exercise. To increase the chance of finding missing annotations while limiting the search space during annotation, potentially high-impact signal words were prioritised. These were drawn from a set of signal phrases that fit the following criteria: (a) more than 10 instances in the corpus, and at least one of: (b) accuracy on positive examples less than 50 % or (c) accuracy on negative examples less than 50 % or (d) below-baseline classification performance. The data from this second pass is in Table 5.13.

5.5.4 *Signal Re-Annotation Observations*

During curation, some observations were made regarding specific signal expressions. In some cases, these observations led to the suggestion of a feature that may help discriminate temporal and non-temporal uses of a certain expression. This section reports those observations.

Previously

TimeBank contains eight instances of the word *previously* that were not annotated as a signal. Of these, all were being used as temporal signals. The word only takes one event or time as its direct argument, which is placed temporally before an event or time that is in focus. For example:

"X reported a third-quarter loss, citing a previously announced capital restructuring program"

Table 5.13 Signal texts that are hard to discriminate; error reduction performance compared to the most common class ("change") is based on a maximum entropy classifier, trained on TimeBank. tp/fn/fp/tn correspond to counts of true and false positives and negatives

Signal	Count	As sig. (%)	Acc. (%)	Change (%)	tp	fn	fp	tn	+ve acc. (%)
for	621	8.2	92.4	8	18	33	14	556	35.3
by	356	5.6	95.2	15	7	13	4	332	35.0
while	39	23.1	79.5	11	1	8	0	30	11.1
from	366	5.2	94.8	0	2	17	2	345	10.5
when	62	85.5	85.5	0	53	0	9	0	100.0
still	35	11.4	88.6	0	0	4	0	31	0.0
already	32	40.6	56.2	−8	1	12	2	17	7.7
at	311	4.8	94.9	−7	2	13	3	293	13.3
as	271	6.6	93.0	−6	3	15	4	249	16.7
over	59	22.0	71.2	−31	7	6	11	35	53.8
since	31	58.1	48.4	−23	12	6	10	3	66.7
then	23	21.7	73.9	−20	0	5	1	17	0.0
earlier	50	12.0	86.0	−17	0	6	1	43	0.0
before	33	93.9	87.9	−100	29	2	2	0	93.5
previously	19	84.2	68.4	−100	13	3	3	0	81.2
former	16	75.0	50.0	−100	5	7	1	3	41.7

In this sentence, the second argument of *previously* is *"announced"*, which is temporally situated before its first argument (*"reported"*). When *previously* occurs at the top of a paragraph, the temporal element that has focus is either document creation time or, if one has been specified in previous discourse, the time currently in focus.

After

Of the nineteen instances of this word not annotated as temporal, only three were actually non-temporal. The cases that were non-temporal were a different sense of the word. The temporal signals are adverbial, with a temporal function. Two non-temporal cases used a positional sense. The last case was in a phrasal verb *to go after*; *"whether we would go after attorney's fees"*.

Throughout

All the cases of *throughout* not marked as signals were not temporal signals. Four were found in the newswire header, which carries meta-information in a controlled language heavily laden with acronyms and jargon and is not prose.

Early

Three of the negative instances of *early* are possibly not correctly annotated; the other 32 negatives are accurate. Of these three, one has a signal use, in part of a longer signal phrase *"as early as"*. The remaining two cases look like temporal signals. However, they are adjectival and only take one argument; there is no comparison, so we cannot say that the argument event is earlier than anything else. For this reason, they are deemed correctly annotated as non-signals.

When

There are 35 annotated and 27 non-annotated occurrences of this phrase. It indicates either an overlap between intervals, or a point relation that matches an interval's start. Twenty-three of the twenty-seven non-annotated occurrences are used as temporal signals. Two of the remaining four are in negated phrases and not used to link an interval pair. for example, *"did not say when the reported attempt occurred"*. The other two are used in context setting phrases, e.g. *"we think he is someone who is capable of rational judgements when it comes to power"* (where *when it comes to* occurs in the sense of *"with regard to"*), which are not temporal in nature.

While

The cases of *while* that have not been annotated as a signal – the majority class, 33 to 6 – are often used in a contrastive sense. This does suggest that the connected events have some overlap, often between statives. For example, *"But while the two Slavic neighbours see themselves as natural partners, their relations since the breakup of the Soviet Union have been bedeviled"*. As two states described in the same sentences are likely to temporally overlap and any events or times outside or bounding these states will be related to the state, it is unlikely that any contribution to TLINK annotation would be made by linking the two states with a "roughly simultaneous" relation; the closest suitable label is TempEval's OVERLAP relation [20].

Example 14 "nor can the government easily back down on promised protection for a privatized company while it proceeds with ..."

The cases of *while* that were not of this sense were easier to annotate. Sometimes it was used as a temporal expression; *"for a while"*. Other times, it was not used in a contrastive sense, but instead modal – see Example 14. The four cases of non-contrastive usage were annotated as temporal signals.

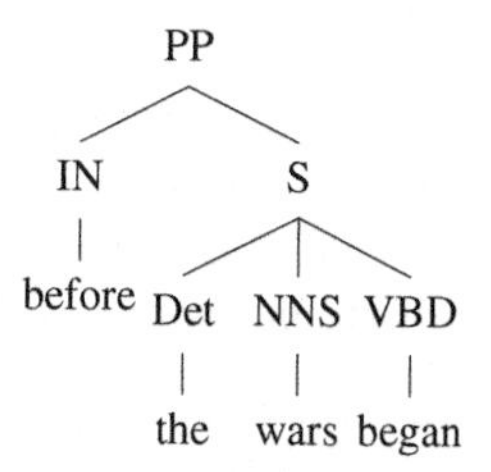

Fig. 5.3 An example of the common syntactic surroundings of a *before* signal

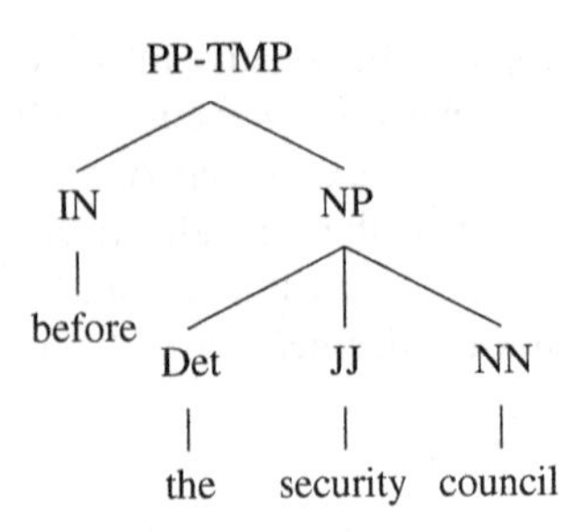

Fig. 5.4 Typical mis-interpretation of a spatial (e.g. non-temporal) usage of *before*. The whole sentence was: *"The procedures are due to go before the Security Council next week."*

Before

Three of the ten negative examples are correctly annotated. They are *before* in the spatial sense of "in front of" (as in *"The procedures are to go before the Security Council next week"*) and also a logical before that does not link instantiated or specific events (*"before taxes"*). The remaining seven unannotated examples of the word are all temporal signals. These directly precede either an NP describing a nominalised event, or directly precede a subordinate clause (e.g. `(IN before, S)` – see Fig. 5.3).

Both cases of *before* that were not temporal signals were parsed and function tagged as if they were.[5] They were given the structure `(PP-TMP, (IN before) ...)` as shown in Fig. 5.4.

Until

All fourteen non-annotated instances of *until* should have been annotated as temporal signals. This word suggests a TimeML IBEFORE relation, unless qualified otherwise by something like "not until" or "at least until".

Already

There were thirteen positive examples of *already*. All of the non-annotated examples had a non-temporal sense as per our description of temporal signals. The word tends to be used for emphasis, but can also suggest a broad "BEFORE DCT" position, which goes without saying for any past and present tensed events. As *already* can be removed without changing the temporal links present in a sentence, no further examples of this were annotated beyond the thirteen present in TimeBank.

Meanwhile

This word tends to refer to a reference or event time introduced earlier in discourse, often from the same sentence. As well as a temporal sense, it can have a contrastive "despite"-like meaning. It is often used to link state-class events, which are difficult to link unless one of their bounds is specific (see Example 15). In this case, it is

[5] Using the PTB trained Stanford Parser and the Blaheta function tagger; see Sect. 5.6.3.1.

not possible to describe the nature of the relation between the start and endpoints of either event interval, and so *meanwhile* suggests some kind of temporal overlap but nothing more. Sometimes *meanwhile* is used with no previous temporal reference. In these cases, the implicit argument is DCT. Five of the ten non-annotated *meanwhile*s were temporal signals.

Example 15 Obama was president. Meanwhile, I was a musician.

Again

This word shows recurrence and is always used for this purpose where it occurs in TimeBank not annotated as a temporal signal. No instances of *"again"* were annotated.

Former

This word indicates a state that persisted before DCT or current speech time and has now finished. Generally the construction that is found is an NP, which contains an optional determiner, followed by *former* and then a substituent NP which may be annotated as an `EVENT` of class STATE. This configuration suggests a `TLINK` that places the event BEFORE the state's utterance.

Example 16 "The San Francisco sewage plant was named in honour of former President Bush."

In Example 16, there is a STATE-class event – *President* – that at one time has applied to the named entity *Bush*. The signal expression *former* indicates that this state terminated BEFORE the time of the sentence's utterance.

Three-quarters of the non-annotated instances of *former* in TimeBank are temporal signals. An example non-temporal occurrence is shown in Fig. 5.5

Recently

Although *recently* is a temporal adverb, it cannot be applied to posterior-tensed verbs (using Reichenbach's tense nomenclature [21]). In the corpus, these are only seen in reported speech or of verbal events that happened before DCT. *Recently* adds a qualitative distance between event and utterance time, but is of reduced use when we can already use tense information.

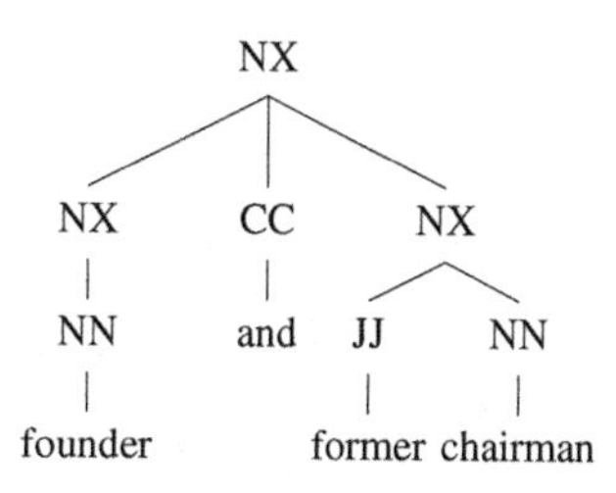

Fig. 5.5 Example of a non-annotated signal (*former*) from TimeBank's `wsj_0778.tml`

The phrase "Until recently" appears awkward when cast as a temporal signal but can be interpreted as "BEFORE DCT", with the interval's endpoint being close to DCT. In this case, recently functions as a temporal expression, not a signal.

Only one of the non-annotated *recently*s in TimeBank is a temporal signal. The exception, *"More recently"*, includes a comparative and is annotated as a TIMEX3; both this phrase and, e.g., *"less recently"* suggest a relation to a previously-mentioned (and in-focus) past event. As a result, we posit that *recently* on its own behaves as an abstract temporal point best annotated as a timex (as seen in the behaviour of *"until recently"* – *until* is the signal here, *recently* a TIMEX3 of value PAST_REF). Structures such as *[comparative] recently* may be interpreted as a qualified temporal signal, as they convey information about the relative ordering of the event that they dominate vent compared with a previously mentioned interval.

5.5.5 *TB-Sig Summary*

Upon examination of the non-annotated instances of words that often occur as a temporal signal (such as *after*) it became evident that TimeBank's signals are under-annotated. As we are certain of some annotation errors in the source data, we revisited the original annotations. A subset of signal words was selected for re-annotation. This set consisted of signals that were ambiguous (occurred temporally close to 50 % of the time) or that we expected, based on informal observations, would yield a number of missed temporal annotations. All temporal instances of these words were re-annotated with TimeML, adding EVENTs, TIMEX3s and TLINKs where necessary to create a signalled TLINK.

A single annotator checked the source documents and annotated 69 extra signals, as well as adding 34 events, 1 temporal expression and 48 extra temporal links. This left 712 SIGNALs that support TLINKs and 780 TLINKs that use a signal, with 54 signals being used by more than one TLINK. No events, timexes or signals were removed.

A summary of frequent candidate signal expressions is given in Table 5.14. The corpus is available via http://derczynski.com/sheffield/. Given this new, curated ground truth for temporal signal annotation, we are now ready to begin approach automatic signal annotation: firstly distinguishing temporal from non-temporal candidate expressions, and then linking signal expressions with the interval annotations that they co-ordinate.

Table 5.14 Frequency of candidate signal expressions in TimeBank and TB-sig. We include counts of how often these occur as signal expressions both before and after manual curation

Expression	Count in corpus	As signal	Proportion as signals (%)	After curation	Proportion (%)
in	1214	161	13.3		
after	72	56	77.8	66	91.7
for	621	52	8.4		
if	65	37	56.9		
when	62	35	56.5	56	90.3
on	344	33	9.6		
until	36	25	69.4	36	100.0
before	33	23	69.7	30	90.9
by	356	20	5.6		
from	366	19	5.2		
since	31	17	54.8	18	58.1
through	69	15	21.7		
as	271	14	5.2		
over	59	14	23.7		
already	32	13	40.6	13	40.6
ended	21	13	61.9		
during	19	13	68.4		
at	311	11	3.5		
previously	19	11	57.9	16	84.2
within	23	8	34.8		
s	10	8	80.0		
later	15	7	46.7		
earlier	50	6	12.0		
while	39	6	15.4	9	23.1
then	23	5	21.7		
once	15	5	33.3		
still	35	4	11.4		
following	15	4	26.7		
meanwhile	14	4	28.6	9	64.3
at the same time	6	4	66.7		
to	1600	3	0.2		
into	63	3	4.8		
follows	4	3	75.0		
subsequently	3	3	100.0		
followed	10	2	20.0	4	40.0
former	16	0	0.0	12	75.0

5.6 Signal Discrimination

The words and phrases that can act as temporal signals do not always convey a temporal relation. Some may indicate possession, or a spatial relation (see Sect. 5.4.4). If we are to automatically annotate signals, we need to develop a method for choosing which words and phrases in a discourse are temporal signals. This task, of finding temporal signal phrases, is called temporal signal **discrimination**.

This section begins with a problem definition and description of the method we adopted to address the problem. An automatic signal discrimination technique is trained using TimeML annotations. Finally, we present results showing automatic accuracy near or above gold-standard corpus IAA.

5.6.1 Problem Definition

The temporal signal discrimination problem is as follows: Given a closed class of signal words or phrases and a discourse annotated with times and events, identify the temporal signals. This task resembles word sense disambiguation [22, 23], in that given a word or phrase that may have multiple senses and its context, we have to determine if the active sense in context is a temporal one.

5.6.2 Method

The approach taken to automatic temporal signal discrimination is a supervised learning one.

We agreed a corpus and a set of words that could occur as signals. Next, we determined a set of feature variables that describe a word in context. After this we described each occurrence of a potential signal phrase in the corpus as a feature vector. Each instance was assigned a binary classification: positive if it is TimeML-annotated as a signal that is associated with a `TLINK`, or negative otherwise. Finally, we trained a classifier with these instances and evaluated its performance.

5.6.3 Discrimination Feature Extraction

As well as surface features from TimeML, syntactic features were used as part of feature extraction for signal discrimination.

5.6.3.1 Parsing and Other Syntactic Annotation

Syntactic information is likely to be of use in the signal discrimination task. Lapata [24] had some measure of success at learning a temporal relation classifier using sentences that contained signals, with syntactic information as a core part of their feature set. Their work used the BLLIP corpus,[6] which contains around 30 million words from Wall Street Journal articles and constituent parses generated by the Charniak parser [25].

To attempt to partially replicate this source information, we parsed the text of the TimeBank corpus. Note that TB-sig and TimeBank differ only in the annotations that they make over text; the actual words in both corpora are the same, and in the same order. To do this, we removed markup from each document and separated the remaining discourse into sentences using the Punkt sentence tokeniser [26], as part of CAVaT preprocessing [11]. Each sentence was then word-tokenised using NLTK's treebank tokeniser.[7] To maintain word alignment consistency with the non-parsed text stored in CAVaT, we needed a parser that accepted external tokenisation. We chose the Stanford parser [27] for generation of constituent parses.

In addition to constituent parses, the BLLIP corpus includes **function tags**. These are optional labels [28] attached to nodes in a constituent tree. Function tags extend a constituent tag by providing additional information about the role it plays in a sentence. They exist in three main groups; syntactic, semantic and topical [29]. Of direct interest to us is the `-TMP` tag, which indicates temporal function. An example of this tag is given in Fig. 5.6, where the first children of an `SBAR-TMP` node comprise a temporal signal.

Early work on function tag assignment in conjunction with the Charniak parser was performed by Blaheta and Charniak [30]. Their approach found that choosing whether or not to assign any tag was a significant and difficult component of the task. Thus, evaluations are split into "with-null" and "no-null" figures, where with-null refers to tag assignment accuracy including the assignment of no tag to untagged constituents and no-null is the proportion of correctly-tagged constituents excluding non-tagged nodes. We refer to no-null performance figures when discussing taggers. The initial Blaheta tagger had an F-measure of 67.8 % on the semantic form/function category, which includes the `TMP` tag.

We would like to use a function tagger with good `TMP` tagging performance. This involved selecting the right tagger. Of these, Musillo [31] simultaneously parsed and tagged text using a Simple Synchrony Parser and an extended tag set. This generated lower results than Blaheta's original attempt though this was improved to provide a marginal increase using input sentences annotated by an SVM tagger. Blaheta's final tagger [32] improved semantic tagging to 83.4 % F-measure, which was comparable to later work in which overall tagging performance increased [33, 34]. As the final Blaheta tagger is freely available and openly distributed, we used this to augment our constituency parser (the Stanford parser [27]).

---

[6] LDC catalogue number LDC2000T43.

[7] See http://www.nltk.org/ for more information on this package.

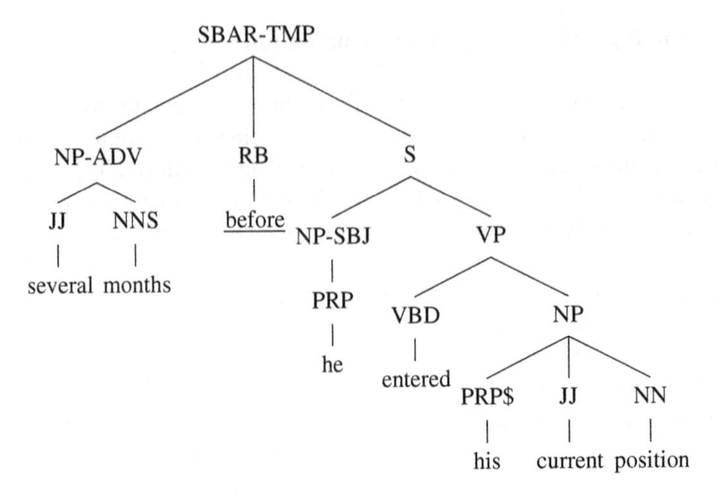

Fig. 5.6 Example of an SBAR-TMP where the first child is a signal qualifier (*several months*) and the second child the signal word itself (*before*)

We only treated as positive examples signals that were associated with a `TLINK`. Signals that only provided information regarding event cardinality, or to subordinate or aspectual links, were ignored. Signals with text not in our closed class of signal words and phrases were ignored.

5.6.3.2 Basic Feature Set

Our initial features were both syntactic and lexical; a list of them is given below. Lexical and TimeML-based features were extracted directly from a CAVaT database constructed from TimeBank [11]. We use NLTK's built-in Maximum Entropy classifier.

a. Part-of-speech from PTB tagset [35]. (`sig_pos`)
b. Function tag from Blaheta tagger; if there is more than one and the set includes TMP, assign TMP, otherwise assign the first listed. (`sig_ftag`)
c. Constituent label and function tag of parent node in parse tree (two features). (`parent_pos, parent_ftag`)
d. Constituent label and function tag of grandparent node in parse tree (two features). (`gparent_pos, gparent_ftag`)
e. Is there any node with the TMP function tag between this token and the parse tree root? (`tmplabel_in_path`)
f. Signal text. (`text`)
g. Text of next token in sentence (if there is one). (`next_token`)
h. Text of previous token in sentence (if there is one). (`previous_token`)
i. Is there a `TIMEX3` in the *n* following tokens? (`timex_in_n_after`)

j. Is there an EVENT in the n following tokens? (event_in_n_after)
k. Is there a TIMEX3 in the n preceding tokens? (timex_in_n_before)
l. Is there an EVENT in the n preceding tokens? (event_in_n_before)
m. The Stanford dependency relation of the candidate word to its parent. ()

In our work, $n = 2$ for the interval proximity features, based on an informed guess after looking at the data. The optimal value, depending on direction of context and type of interval (event vs. timex) search for, is left to future work.

There are 102 entries in our closed class of signal words/phrases; this set is kept constant throughout all experiments. In TimeBank there are 7 014 mentions of the members of this set, including both temporal and non-temporal mentions.

5.6.3.3 Extended Feature Set

Curation of signals, as detailed in Sect. 5.5, led to some direct observations about specific signal words. These observations in some cases suggested specific sources of signal discrimination information thar could potentially be translated to features. From the observations above, the new features that could be added were:

n. Flag to see if signal text is in a verb group (*before, after*) (in_verb_group)
o. Flag to see if a token at the top of a paragraph (*previously*)
p. Flags to see if the preceding or following word(s) are part of a verb group (*after*) (following / preceding_in_verb_group)
q. What is the highest-level subtree that begins at the next token (*before*) (following_subtree)
r. What is the highest-level subtree that ends at the preceding token (preceding_subtree)
s. PoS of the next token and previous token (*before*, *after*) (following/preceding_pos)
t. PoS of the next event within n tokens (*before, former*) (next_event_pos)
u. Type (TimeML class) of the next event within n tokens (*former*, *meanwhile*) (next_event_class)
v. TimeML Tense and aspect of the next event within n tokens (*already*) (next_event_tense / aspect)
w. NP begins at next token? (*former*) (np_next)
x. Is the preceding token a comparative, i.e., is it one of JJR or RBR? (*recently*) (preceding_comparative)

All of these were implemented and added as features, except the paragraph-top feature (due to a lack of a reliable document segmentation tool). In addition, we removed some noisy features that seemed to be causing overfitting within our sparse data set; the offset of the word within its sentence and the preceding & following token texts. We used the full constituent tag of subtrees for the preceding_subtree and following_subtree features, including.

Table 5.15 Comparison of the effect that decomposing values of the preceding_subtree and following_subtree features has, using our extended feature set and TimeBank data. Error reduction compared to classifier MCC baseline

Features	NBayes	MaxEnt	ID3
Full subtree labels	−1.32	19.4	**25.4**
Just constituent tag	−2.31	19.7	**21.6**
Separate constituent and function tag	−4.28	19.9	**24.2**

5.6.3.4 Multivalent Tags

In a minority of cases, constituents and terminals were assigned multiple function tags. For example, values such as PRD-TPC-NOM or TMP-SBJ would be appended. Noticing that these instances were assigned high weights by a Naïve Bayes classifier, we measured error reduction on multiple variations of subtree tag feature representations. Results are shown in Table 5.15. It was found that reducing data sparsity by providing two separate features per subtree (for constituent tag and function tag) provided best overall performance for MaxEnt discriminators, but ID3 benefited most from the feature extraction that gave the sparsest values – full subtree labels.

5.6.3.5 Choice of Learning Algorithm

Signal discrimination is a binary classification problem: is a given word or phrase a temporal signal or not? We have constrained the set of words we attempt to classify by defining a closed class of signal words and described a set of features with which we will represent candidate words and context. We now need to choose a binary classification algorithm. We use a Naïve Bayes classifier, decision trees, a maximum-entropy classifier and adaptive boosting.

For rapid learning and quick feedback, we worked with the Naïve Bayes classifier. Naïve Bayes models are computationally cheap to learn. Its inductive bias includes the independence assumption – that all features are independent from each other. This is not true in our case, given the heavily interdependent nature of most of our features: well-formed syntactic structures are inherently constrained by grammar and the values of many of our features depend on syntax at multiple places in the same sentence or paragraph. For example, the parts of speech of any given token has some bearing on the part of speech of the following one, and these are again not independent of the parse tree of the sentence in which they occur. We also use a decision tree classifiers, which do not have this particular bias and are computationally quick to learn, but do not always cope well with noise. ID3 and C4.5 types are used. C4.5 attempts to deal with noise in training data by performing pruning on the tree after construction [36].

We also evaluate performance of our feature set with a maximum entropy classifier. This regression-based model assumes low collinearity between features, which is a less constraining assumption than that of the Naïve Bayes classifier, though problems may arise if we use highly-correlated features. Finally, we use adaptive boosting with decision stumps [37, 38], which is constrained to binary classification and can yield high-performance results. Adaptive boosting reduces the impact of the typically computationally intensive SVM-learning process and typically displays little overfitting, which is helpful with smaller datasets such as ours.

Performance was improved by removing features that have a high number of values (for example, the text of the token after a signal). We suspect this is due to them leading to overfitting.

5.6.4 Discrimination Evaluation

We have described how we trained a classifier using cross-validation. We evaluated performance using a held-out evaluation set, and determined scores by counting correct classifications and measuring both percentage of correctly classified instances and also the error-reduction compared to a baseline.

5.6.4.1 Baselines

To evaluate the performance of our approaches, it is useful to describe some simple annotation methods as baselines. A summary of our baselines is given in Table 5.16 and we explain each of them below.

One simple baseline is to find the most common classification and assign this to all instances. In our corpus, instances of phrases from our list of potential signals are used non-temporally nearly all the time (out of 6 091 instances of potential signal phrases, only 688 are annotated as being temporal signals in TimeBank – 11.3 %) and so our most common case is to classify everything as not being a temporal signal, regardless of the signal text.

We also use baselines that mark all words found in the signal phrase list as temporal signals if they have a part-of-speech tag of RB or IN, according to NLTK's built-in

Table 5.16 Performance of four constituent-tag based baselines over TimeBank

Baseline	Accuracy (%)	Accuracy on positives (%)
Most common class	86.7	0.0
Baseline: Part-of-speech is IN	25.6	**81.2**
Baseline: Part-of-speech is WRB	86.9	5.77
Baseline: Parent is SBAR-TMP	**87.0**	9.88
Baseline: Parent function is -TMP	84.5	72.7

maximum entropy tagger. Values are quoted for overall classification accuracy, as well as accuracy on positive examples (the minority of our training data).

Most Common Class

The training set is confined to just signal annotations in TimeBank/TB-sig, that are also in the closed class of signal expressions detailed above in Table 5.9. This introduces an inherent performance cap to the overall approach, but assumes no knowledge of whichever corpus is being used as the evaluation set. Of 4 576 training instances, 3 969 are negative (non-temporal) and 607 are positive (having a temporal meaning). The most-common-class is negative and if we assign this label to all mentions of members of the set, classifier accuracy is 86.7 % but no signals are identified (giving an effective F1 of zero if we imagine this as a signal recognition task); not a very informative baseline.

Class Member and Signal Word Tag

Of all leaf labels, IN and WRB have the highest proportion of signals (Table 5.10). To this end, we have two simple baselines, where we count a word as a temporal signal if its constituent tag is IN or WRB and it is found in the closed class of signals. Performance for these is given in Table 5.16. For IN, we have 25.6 % overall accuracy, correctly identifying text that is a temporal signal 81.2 % of the time. For WRB, we achieve 86.9 % accuracy, but only 5.77 % on the positive examples.

Parent Is SBAR-TMP

As mentioned in Sect. 5.4.2.1, one might expect an a SBAR-TMP subtree to begin with a temporal signal and also contain one of the signal's arguments (see also Fig. 5.6). As we can use our closed class of signal words to differentiate signal head, signal qualifier and event/timex argument, we can look for leaves where the parent is SBAR with TMP in its function tags. This is our SBAR-TMP baseline, that performs at 87.0% accuracy overall, with 9.88 % on positives – better than WRB, but still poor.

Parent Has Temporal Function

Limiting ourselves to just signals in subtrees labelled SBAR may be a short-sighted manoeuvre. We added a baseline that labels signal candidates as temporal if their parent has a temporal function label. This baseline achieves classification accuracy of 84.5 % and a 72.7 % accuracy on the positive examples; see Table 5.16.

Table 5.17 Signal discrimination performance on the plain TimeBank corpus. Error reduction is measured relative to the "parent has temporal function" baseline. Evaluated with 5-fold cross validation and 1 000 iterations of adaptive boosting

Measure	Accuracy	Accuracy (+ve)	Error reduction	Error reduction (+ve)
Naïve Bayes	88.6	**78.4**	26.5	**20.9**
Maximum Entropy	89.5	56.0	32.3	−61.2
ID3	90.5	65.6	38.7	−26.0
C4.5	90.4	60.1	38.1	−46.2
AdaBoost	**90.7**	59.8	**40.0**	−47.3

5.6.4.2 Performance

With our original feature set and based on pre-curation data (e.g. TimeBank v1.2), we achieved a 40 % error reduction in signal discrimination relative to a competitive baseline, as seen in Table 5.17. For the general annotation task, naïve Bayes performed best, with good error reduction overall (26.5 %) and a similar improvement in recognition of positive examples (20.9 %), something that other classifiers did not perform so well with.

With the original feature set, models learned over TB-sig data performed as shown in Table 5.18. Performance using the extended feature set is detailed in Table 5.19, again based on TB-sig.

Our extra annotations introduce new signal instances for the extra terms that we have annotated, reducing the baseline to 85.2 % accuracy (677 positives, compared to 607 before re-annotation) from 86.7 % before – see Table 5.18. Performance using TB-sig is overall better (compared to Table 5.17), which we attribute to having a better-stated problem and less misleading data. Error reduction rate is now over 40 %, with overall accuracy just under 92 % and up to 75 % on the positive examples. This is better than performance on the original TimeBank data and comparable to the IAA figure of 0.77 for TimeBank's initial `SIGNAL` annotation. C4.5 performs particularly well, reaching near-highest error reduction rate and good accuracy on positive examples.

The extended feature set, however, does not improve performance in the majority of cases, despite having been generated as part of a rational investigation. Analysis and further work is required to improve upon these signal discrimination results.

Table 5.18 Signal discrimination performance on the curated corpus. Error reduction is measured relative to performance. Results are for 5-fold cross validation. Adaptive boosting used 1 000 iterations

Measure	Accuracy	Acc. (+ve)	Error reduc.	Error reduc. (+ve)
Most common class	85.2	0	n/a	n/a
Baseline: IN	25.4	77.1	–	–
Baseline: RB	86.3	8.3	–	–
Baseline: SBAR-TMP	86.1	10.8	–	–
Baseline: Temporal parent	84.5	70.0	–	–
Simple features				
Naïve Bayes	89.3	**78.7**	31.0	29.0
Maximum Entropy	88.2	51.3	23.9	−62.3
ID3	91.7	69.6	46.5	−1.3
C4.5	**92.1**	73.0	49.0	10.0
AdaBoost	91.9	70.5	47.7	1.7
Extended features				
Naïve Bayes	87.0	**81.4**	16.1	38.0
Maximum Entropy	88.1	50.1	23.2	−66.3
ID3	91.1	68.7	42.6	−4.3
C4.5	91.7	75.0	46.5	16.7
AdaBoost	**91.8**	69.3	47.1	−2.3

Table 5.19 Signal discrimination performance on the TimeBank corpus, with an extended feature set. Error reduction is measured relative to most-common-class ("not a signal") performance. Evaluated with 5-fold cross validation and 1 000 iterations of adaptive boosting

Measure	Accuracy	Accuracy (+ve)	Error reduction
Extended features			
Naïve Bayes	86.1	80.9	−4.28
Maximum Entropy	89.4	55.4	19.9
ID3	89.9	59.8	24.2
C4.5	90.4	60.8	27.8
AdaBoost	**90.6**	**59.3**	**29.3**

5.6.4.3 Useful Features

A sample post-classification analysis of feature weights – using TB-sig and the extended feature set – is presented in Table 5.20, taken from the last of five cross-validation passes. This is from the construction of a model using the whole signal-labelled corpus with a naïve Bayes classifier. The text of the signal is a particularly strong indicator for some of the features that occur much more often as temporal signals than not. We can also see that wh-adverb signals and wh-adverb phrases that contain the candidate signal expression are strong indicators of temporal meanings (features signal_label, parent_label and ending_subtree_label); this may be because of words such as *when* having only temporal senses. A timex or a past-tensed event occurring after the signal is also an indicator of it being temporal (timex_in_2_after). When the parent constituent or the largest constituent beginning at this point has a temporal function, then a candidate word is more likely to be temporal (parent_function, starting_subtree_function). The -TMP function tag helps to indicate a temporal signal when it dominates the candidate signal word (tmpfunction_in_path). Being followed by a dollar amount suggests that a candidate is not temporal (following_label = $) – for example, in a non-temporal use, *"Shares closed at $ 50"*; the high weight of this attribute-value pair is likely influenced by the high proportion of financial reporting in TimeBank, which takes a significant part of its text from the Wall Street Journal.

Words and phrases that are within a syntactical structure that has a spatial function (e.g. -LOC) contra-indicate a temporal meaning. This is aligned with the observation that members of our class of signal words often have both temporal and spatial meanings. Further, an adjacent structure with a spatial function (-EXT or -LOC) suggests a temporal function in a candidate word. This suggests collocation based approaches may not correctly discriminate temporal and non-temporal signals; syntactic parsing is required, in order to detect these functional nuances. Having NX (indicating the head of a complex NP) as a parent at can indicate a signal; this could be in cases where we have a signal before a nominalised event, such as in *"before the explosion"*. Finally, preceding a verb may be an indication of a temporal signal; this reflects the signal's adverbial nature.

5.6.5 Discrimination on Unseen Data

Up to this point, evaluation has used cross-validation over TimeBank. Our error analysis led to the inclusion of features based on the data that is also part of the evaluation set. To check performance on previously unseen data, a further experiment was performed is as follows. We trained a signal discriminator and associator based on all of TimeBank + the extra signal annotations. The closed class is increased to include all phrases marked as signals in TimeBank. This way, TimeBank is only the training data.

Table 5.20 Sample features useful for signal discrimination, based on our curated TimeBank data, TB-sig

Feature	Value	Indication	Weight
text	until	True	131.5
text	before	True	70.0
text	after	True	56.9
signal_label	WRB	True	49.6
parent_label	WHADVP	True	49.5
ending_subtree_label	WRB	True	48.5
text	when	True	48.3
text	previously	True	26.2
text	former	True	15.4
grandparent_label	SBAR	True	13.9
text	during	True	11.5
following_subtree_function	-LGS	False	9.7
text	meanwhile	True	9.6
timex_in_2_after	True	True	9.0
text	since	True	7.6
preceding_subtree_label	S	True	7.2
starting_subtree_function	-LOC	False	7.1
following_label	$	False	7.0
starting_subtree_label	SBAR	True	6.6
parent_function	-LOC	False	6.4
following_subtree_label	VBN	True	6.3
starting_subtree_function	-TMP	True	6.2
following_label	PRP	True	6.1
grandparent_label	NX	True	5.7
starting_subtree_label	NX	True	5.7
preceding_label	JJS	True	5.6
following_subtree_label	VB	True	5.6
text	thereafter	True	5.6
next_event_tense	PAST	True	5.4
parent_function	-TMP	True	5.3
parent_label	SBAR	True	5.3
text	later	True	4.9
tmpfunction_in_path	True	True	4.1
preceding_subtree_function	-LGS	True	4.1
preceding_subtree_function	-EXT	True	4.1
following_subtree_function	-PRD	True	4.1
starting_subtree_function	-TPC	True	4.1
grandparent_label	SINV	True	4.1
following_subtree_label	.	False	4.0
following_label	.	False	4.0

Table 5.21 Characteristics of the N45 section of the AQUAINT TimeML corpus, before and after signal curation

Feature	Pre-curation	Post-curation
Documents	15	
Tokens	7099	
Signals	96	114
TLINKs	1048	1062
Events	1060	1060
Timexes	154	156

Table 5.22 Performance of a TB-sig trained signal discriminator on unseen data

Method	Accuracy (%)	Precision (%)	Recall/acc. on positives (%)
Parent -TMP baseline	84.5	–	70.0
MaxEnt model	93.6	83.0	78.3

As the final model was developed based partially on observations of TimeBank, it is not suitable to evaluate the final model on this corpus also. A previously unseen set, taken from the AQUAINT corpus (Sect. A.2.2), now forms the evaluation set. The N45 section of the AQUAINT corpus was curated to verify its signal annotations, and then signal discrimination was evaluated over this subcorpus based on a model trained on the entirety of TB-sig. The relevant statistics regarding this evaluation corpus are presented in Table 5.21.

Signal discrimination is measured in two ways. Firstly, classification accuracy shows how many of the candidate signal words were correctly labelled as signals or not-signals. Secondly, the overall performance of the association approach at annotating signals in any given document is described in terms of precision and recall. This takes into account how well the entire approach described above (including the signal words list described in Table 5.9, but not also including those found in TimeBank) does when given the task of identifying temporal signals in an arbitrary text. The augmented AQ/N45 annotations form the gold standard. The "parent has temporal function" baseline (Sect. 5.6.4.1) is used for comparison. Results are presented in Table 5.22. This compares well with the performance on (seen) TB-sig data (Table 5.17).

5.6.6 Summary

In this section, we have explored the task of signal discrimination. We discovered that TimeBank's signal annotations are incomplete. To remedy this, we have proposed augmentations to the TimeML annotation standards and re-annotated a portion of the corpus. We have also defined a set of features that can describe a temporal signal in context and constrained our search space to just words and phrases in a closed class of signal words. As a result, we have been able to train a classifier to detect temporal signals at near-IAA accuracy.

5.7 Signal Association

Temporal signals connect one or more interval pairs and describe the nature of the temporal relation between the pair. This section describes an investigation into how to find the arguments of a temporal signal, thus associating the two arguments. We refer to this task as **signal association**.

In order to fully annotate temporal signals, we need to determine which arguments they co-ordinate. To this end, the task of determining which times or events are coordinated by a temporal signal is examined as the subject of this section.

5.7.1 Problem Definition

When performing temporal annotation, one needs to identify events and times and can then connect them with temporal links, perhaps using an associated signal. In fact, every time that a temporal signal is annotated, there must be a temporal link present. The signal association problem is: Given text with signal, event and timex annotations, determine which pair of events/times are associated by each signal phrase.

5.7.2 Method

A supervised learning approach is taken to finding which intervals a given signal co-ordinates. TB-Sig is used as the dataset for feature extraction. Two approaches are explored, detailed below. These use a largely common feature set, extracting a number of features for each interval considered and a further set of features describing the signal.

To generate training data given a signal, we will describe events and timexes within the scope of that signal using our feature set. Although any two intervals in a document could be linked by a given signal, the number of intervals or interval pairings one must search through could be large if the entire document is used as potential signal scope. For this reason, scope must be constrained, at a possible performance loss. Given candid examination of the signals in the corpus, the scope of the signal is taken to be the signal's sentence and also enough previous sentences to include at least two intervals, as well as a DCT timex if present. We are attempting to determine which intervals are associated with the signal.

The goal is to learn a binary function, that can indicate whether or not an association supporting a TLINK exists in a given situation. A TLINK associates two intervals (timex or event) and may specify the type of temporal relation between them. We have tried two approaches to this signal association task; one where we examine ⟨interval, signal⟩ tuples and another where we examine ⟨interval-pair, signal⟩ tuples. The gold standard corpus, TimeBank, provides the positive examples. For each signal, there may be up to five valid TLINKs, each shown as an interval pair (see earlier Table 5.8).

For the single interval approach, we train a binary classifier to learn if an interval and signal are linked and then choose the two best candidate intervals for a signal, using classifier confidence to rank similarly-classified intervals. For the interval pair approach, for each signal we examine possible combinations of intervals and create a vector of features based on relations between the intervals and the given signal.

5.7.2.1 Single Interval Approach

In this section, we describe a signal association approach where individual intervals are ranked by their relation to the signal and the top two intervals are deemed to be associated.

Positive training examples came from intervals associated in a gold standard annotation. Negative training examples were taken to be all temporal intervals in the same sentence as the signal that were not associated with the signal. We used cross-validation to learn classifiers and recorded the prediction and confidence of the classifier for each entry in the evaluation fold. After this, for each signal, a list of candidate intervals was determined. The two intervals related to the signal were those classified as related with highest classifier confidence, or if fewer than two positive classifications were made, up to two are taken from lowest-confidence unrelated classifications. That is, for each signal, intervals are ranked in descending order of confidence; the goal is to find the two most likely intervals, and associate them in a TLINK backed by the given signal. Priority is established in this order:

1. High-confidence and classified as related
2. Low-confidence and classified as related
3. Low-confidence and classified as unrelated
4. High-confidence and classified as unrelated

The top two are then associated with a signal. This approach is limited to only detect one pair of intervals per signal.

5.7.2.2 Interval Pair Approach

In contrast to our previous approach, we tried to identify whole ⟨interval-pair, signal⟩ 3-tuples as either a signalled `TLINK` or not. This produced a majority of negative examples. We instead only considered intervals where both arguments fell inside a sliding window of sentences, to reduce the heavy skew in training data. A boolean feature describing whether the intervals were in the same sentence was added to our set, as well as two sets of interval-signal relation features and general signal features as described earlier.

5.7.2.3 Surface and Constituent-Parse Features

For the signal association tasks, we used the following surface and constituent-parse features as input to a binary classifier. Constituent parse information comes from running the Stanford Parser [27] over discourse sentences, the bounds of which are determined using the Punkt tokeniser [26] implementation in NLTK. The features describe a single interval/signal pair. We use the same definition of syntactic dominance as [24]; that is, an interval (e.g. event or timex) is syntactically dominated by a signal if the interval's annotated lexicalisation is found within a parse subtree where the first (leftmost) word of the parse subtree is the signal. Dominance features are included based on their success in signal linking in [24], where dominance was described as the V_L feature.

- Is this interval the textually nearest after the signal?
- Is this interval the textually nearest before the signal?
- Does the signal syntactically dominate the interval?
- Signal text (lower case)
- Signal part of speech
- Token distance of interval from signal
- Interval/signal textual order
- Is there a comma between the interval and signal?
- Is the interval in the same sentence as the signal?
- Is the interval DCT or a DCT reference?
- Interval type (TimeML `EVENT` class or `TIMEX3` type), total 11 values
- If an event, its TimeML-annotated tense

5.7.2.4 Dependency Parse Features

We use the Stanford dependency parser [39] to return dependency graphs of our PoS-tagged, parsed and function labelled sentences. By default, the dependency parser ignores some words that we consider to be signal words, moving information about removed words in relationships. We configured it to never ignore words. The features that we extracted from sentence dependency parses were:

- Length of path from interval to root
- Is the signal a child of the interval?
- Is the signal a direct parent of the interval?
- Is the interval the tree root? (e.g., the head event/time)
- Is the interval directly related to the signal with an `advmod` or `advcl` relation?
- Does the interval modify the root directly? (e.g., is the interval a direct ancestor of the root, regardless of relation type)
- Does the signal modify the interval directly? (e.g., is the signal a direct ancestor of the interval)
- What relation does the interval have to its parent?
- If the signal is a child of the interval, what is the relationship type?

5.7.3 Dataset

Examining some of the instances of temporal relations in TimeBank which have an attached signal, there were often clear syntactic relations between signals and their arguments (which are also the temporal relation's arguments). Almost all signals co-ordinated two intervals in the same sentence as the signal (Table 5.23). In the cases where they did not, one of three situations prevailed. Firstly, the signal was the first token in the sentence and the argument outside of the sentence was either referenced by a temporal pronoun (as in e.g. *"After **that**, the situation improved."*). Secondly, one argument is an event or time that has remained the temporal focus in discourse at the point where the signal is found, even after new sentences have been introduced. Thirdly, the signal will relate DCT with an interval in its sentence.

5.7.3.1 Closure

Some supervised approaches that deal with temporal relations chose to use closure to generate extra training data. We have deliberately chosen not to include temporal

Table 5.23 Distribution of sentence distance between intervals linked by a signal, for TB-sig. A special case is made for those that link to document creation time or one of its co-referents, as it often persists as a reference point through the length of a discourse

Distance	Count
DCT	40
0	682
1	43
2	16
3	3
4	3
5+	0

links generated through closure [40] in our examples. Temporal closure typically generates more links than were in the original annotation by at least an order of magnitude. The generated links tend to be between intervals not directly related in text – e.g. lacking textual proximity or clear discourse relations. As with many binary classification models, the negative examples that enable our classifiers to learn the most precise decision boundaries are those that closely resemble positives. Entities only linked through a chain of four or five annotated `TLINK`s, with low textual or syntactic proximity, will not be in this set. We do however use windowing approaches to permit some of these wide-ranging negative examples into the training.

5.7.3.2 Detecting Document Creation Time

Document creation time (**DCT**) refers to the instant at which a discourse was created. In the case of newswire articles this is often included in the article metadata, or as a deictic temporal expression at the beginning of the first sentence, which describes day and month (e.g. *"KABUL, August 21 – ..."*). Other times, it may be possible to extract this date automatically [41]. The document creation time persists throughout a discourse as an antecedent temporal point that may be referred to by temporal expressions or, in some cases, signals. As we have seen some signals that work like this (e.g. *afterwards*), it may be useful to include a boolean feature indicating whether or not a timex represents DCT.

TimeML-annotated data is used to determine whether a given timex is DCT or DCT-equivalent. Our algorithm is as follows, given a candidate `TIMEX3` element:

1. if `functionInDocument` = `CREATION_TIME` ⇒ return **true**
2. if `functionInDocument` = `PUBLICATION_TIME` ⇒ return **true**
3. `most-frequent-anchor` ← the most frequent non-null value of `anchor TimeID` in this document's `TIMEX3` annotations
4. if `sentence-number` < j and `timex_id` = `most_frequent_anchor` ⇒ return **true**
5. else return **false**

That is, we first look for explicit annotation markers that declare this timex to be a creation time reference. Failing that, if the timex is near the beginning of the document and also the timex most-often used as an anchoring point for other timexes, we mark it as DCT-referring. With $j = 2$, this heuristic is accurate for all of TimeBank.

5.7.4 *Automatic Association Evaluation*

As both approaches rely on a binary classifier, the first evaluation measure given is classifier accuracy. This shows the proportion of accurate binary decisions made by the classifier based on model learned from training data. The error reduction that the

Table 5.24 Performance at the signal:interval association task, with 5-fold cross validation. The classifier performance baseline is most-common-class, which was 64.1 % not-related for TimeBank and 64.0 % not-related for the signal-augmented version

Corpus	Classifier	Accuracy	Err. reduc	Full (%)	Partial (%)	Failure (%)
TimeBank	MaxEnt	**85.2**	58.7	**64.2**	34.5	1.25
	NBayes	82.5	51.1	57.2	41.2	1.53
	ID3	78.4	39.8	42.1	52.1	5.85
TB-sig	MaxEnt	**84.8**	57.9	**61.5**	37.6	0.897
	NBayes	82.2	50.5	56.3	41.9	1.79
	ID3	79.6	43.4	40.9	54.4	4.74

classifier's model provides over a most-common-class baseline is also given. The single-interval approach and interval-pair approaches are structurally different and can be further evaluated in separate ways, which are detailed below, as well as results.

5.7.4.1 Single-Interval

We recognised three possible states of signal annotation. A **full match** occurs when both signal arguments are correctly found, when just one argument is correct we have a **partial match** and when both associated arguments are incorrect there is a **failure**. Results of classifier performance and signal annotation success can be found in Table 5.24. Full matches are the only cases we should consider as successes; anything else is not correct, though partial successes (where one argument is correctly associated) are shown to give insight into how problematic the non-full matches were. As can be seen from the data, even in cases where there was not a full argument match, it was almost always the case that at least one interval was correctly associated – that is to say, partial matches were orders of magnitude more common than failures.

5.7.4.2 Interval-Pair

Results for the interval-pair:signal approach are given in Table 5.25. The “Acc (+ve)” column represents the classifier accuracy on examples labelled as positive in the gold standard, as opposed to the proportion of the instances labelled as positive that were matched the gold standard annotations. The best classifiers are those that achieve a high error reduction while maintaining good classification accuracy on positive examples.

For most Naïve Bayes classifier results, there were was a low false negative and a high true positive rate, but also an overbearing false positive rate. For example, with $n = 2$ there were 1371 true positives and only 65 false negatives, which is good, but 4513 false positives, meaning that the classifier output was not particularly useful. Less than one quarter of interval-pair:signal associations would be accurate.

Table 5.25 Performance at the signal:interval-pair association task, with 5-fold cross validation. The baseline is most-common-class, which was "no link" in all cases. The sentence window for negative examples is the signal's sentence plus the n prior sentences

Corpus	Classifier	Accuracy	Err. reduction (%)	Acc. (+ve)
TimeBank n = 0, baseline 89.6	NBayes	94.0	41.8	91.4
	ID3	97.7	77.3	84.7
	MaxEnt	92.5	28.0	43.7
TimeBank n = 1, baseline 96.6	NBayes	93.6	−89.4	93.9
	ID3	99.3	**79.9**	84.0
	MaxEnt	97.1	13.9	43.6
TimeBank n = 2, baseline 98.3	NBayes	94.7	−219	**95.5**
	ID3	**99.4**	62.1	68.7
	MaxEnt	84.9	-804	39.3
TB-sig n = 0, baseline 89.7	NBayes	94.1	42.8	90.8
	ID3	97.4	74.8	84.8
	MaxEnt	92.2	23.6	41.6
TB-sig n = 1, baseline 96.7	NBayes	93.4	−100	93.2
	ID3	99.3	**78.0**	83.5i
	MaxEnt	97.1	12.3	44.5
TB-sig n = 2, baseline 98.4	NBayes	94.7	−229	**94.7**
	ID3	**99.1**	42.7	46.8
	MaxEnt	84.9	−832	38.8

Table 5.26 Confusion matrix for signal association performance with a MaxEnt classifier on TimeBank with a window including the signal sentence and two preceding ones

	Prediction	
Class	True	False
True	564	872
False	12,110	72,192

Table 5.26 shows the confusion matrix of the worst-performing attempt. It detects a large number of false positives.

Using windowing for candidate interval selection with $n = 2$, 0.38 % of signal arguments lie out of the window (see Table 5.27) and are therefore not correctly associable with this approach – an acceptably small amount. With $n = 0$, this unassociable proportion rises to 4.13 %. We found that increasing n led to worse classifier performance and a value of $n = 1$ provided a good trade-off.

Table 5.27 Distribution of sentence distance between intervals and signal that links them. A special case is made for those that link to document creation time or one of its co-referents, as in Table 5.23

Distance	Count
DCT	41
0	1468
1	43
2	16
3	3
4	3
5+	0

Performance is worst with $n = 2$. We can achieve a good classification accuracy on a test set that includes cross-sentence links even if we only consider same-sentence intervals for the generation of negative examples (i.e. $n = 0$). We can also see that decision trees, which do not follow the independence assumption, perform consistently well, although do worse as n increases.

5.7.4.3 Evaluating on Previously Unseen Data

To test association on its own, a classifier is trained on TB-sig and evaluated on the augmented AQ/N45 data (a TimeML subcorpus introduced in Sect. 5.6.5). The interval pair annotation method is used, as it performs best on prior TimeML data (Sect. 5.7.4.2). The results are shown in Table 5.28.

This is satisfactory performance, with a strong error reduction of 58 % beyond the baseline.

5.7.5 Association Summary

Our aim was to find a method of automatically associating a temporal signal with a pair of intervals, given a partially annotated text. We tried two approaches. The first ranked ⟨interval, signal⟩ tuples and treated the top two as linked. The second treated ⟨interval-pair, signal⟩ tuples as atomic units.

Table 5.28 Performing of a TB-sig trained signal associator on unseen data

Method	Accuracy (%)	Error reduction (%)	Acc. on positives (%)
Most common class (not related)	91.96	–	0.00
ID model (n = 1)	96.60	57.72	84.93

It is important to achieve a good error reduction rate and also to have good predictive accuracy on positive examples. Both of these metrics need to have high values for a classifier to be useful in annotation. We found that although the ranked single-interval approach achieved decent results, treating interval pairs as atomic units worked better. We achieved 78.0 % error reduction over the most-common-class baseline, at 96.7 % predictive accuracy and 83.5 % accuracy on the positive examples.

5.8 Overall Signal Annotation

The overall motivation for signal extraction is to improve automatic temporal relation typing. We have independently determined that signals are useful for TLINK typing (Sect. 5.3) and that we can extract and associate signals automatically (Sects. 5.6 and 5.7). To show that automatic extraction is useful in support of the relation typing task, we took a gold-standard TimeML corpus (the AQUAINT TimeML corpus) and removed all its signal annotations. Performance of an automatic TLINK labeller was then compared when there are no signal annotations and when signal annotations have been automatically added using the above methods.

The same unseen corpus (a signal-augmented version of the N45 section of AQUAINT TimeML corpus) was used for evaluation of discrimination and association, as introduced in Sect. 5.6.5.

5.8.1 Joint Annotation Task

To measure combined performance, the signal annotations suggested in the discrimination step are used as the basis for association. Note that because the set of TLINKs identified in a document's annotation may not be a temporal closure of that document (see Sect. 3.3.2), it is possible to correctly detect a pair of events that are in fact linked via a signal but for the TLINK not to be present in the gold standard. For this reason, the performance scores are minimums. We hypothesise that despite a lack of guidance regarding which TLINKs must be defined in order to create a complete or valid TimeML annotation, annotators are likely to add explicit TLINK annotations where the temporal relation is suggested explicitly (e.g. with a signal). Therefore the number of unannotated signalled TLINKs should be small.

The corpus used was the augmented N45 dataset, stripped of TLINK and SIGNAL annotations (leaving TIMEX3s and EVENTs). The method was to first attempt automatic signal discrimination over the corpus (training on all of TB-sig using the basic feature set), and then perform automatic signal association (using the interval-pair approach). The resulting SIGNAL and TLINK annotations were then compared to the augmented N45 annotations.

Table 5.29 Details of the joint approach to signal annotation. Although the augmented N45 corpus only contained 136 signals, our approach found 424. This table breaks down that 424

Signal/TLINK associations	Count	Proportion (%)
In N45	136	–
Found	336	–
Found, both args in N45	88	26.2
Signal in N45, new TLINK assoc	216	64.3
Found based on new signals	32	9.5

Results are summarised in Table 5.29. In total, compared to the 136 signalled TLINKs in the augmented AQ/N45 data, 336 interval pairs (e.g. TLINK suggestions) were suggested based on the automatically annotated signals. A total of 64.7 % of the 136 TLINKs were found correctly automatically. Only 26.2 % of associated interval pairs (88 out of 424) were found in the gold standard; 248 were not there. A minority of 9.5 % (32) of pairs found were based on signals not in the gold standard. This leaves 64.3 % (216) automatically generated instances of signal associations with interval pairs not mentioned in the gold standard.

Upon manual inspection, many of these false positives based on existing signals appear to be supported in the text, but are not annotated in the gold standard, which in many cases contains only a minimal annotation, and certainly never constitutes a closure. Take the following cases, for example, taken from NYT19990505.0443.tml in the signal-augmented corpus and edited slightly for brevity:

Example 17 A jogger <EVENT eid="e64">*observed*</EVENT> Kopp's car <SIGNAL sid="s7">*at*</SIGNAL><TIMEX3 tid="t10">*6a.m.*</TIMEX3>near Slepian's home <TIMEX3 tid="t11">*10 days*</TIMEX3> <SIGNAL sid="s8">*before*</SIGNAL> the <EVENT eid="e65">*murder*</EVENT>, and, <EVENT eid="e66">*curious*</EVENT> why a stranger would be <EVENT eid="e67">*parked*</EVENT> there so early, <EVENT eid="e68">*wrote*</EVENT> down the license plate number.

In this section, our approach found the links listed in Table 5.30 (in this example, event eids and instance eiids have a 1:1 mapping, so ei65 corresponds to event e65).

Table 5.30 Sample signals and arguments found in N45

Signal ID	Argument 1	textbfArgument 2	In GS?
s8	ei64	ei65	Yes
s8	ei65	ei66	No
s8	ei65	ei67	No
s8	ei65	ei68	No
s8	ei65	t1	No
s8	ei65	t11	Yes

Many of the links suggested but not annotated are in fact correct from the text. For example; signal s8 (*before*) is said to describe the temporal relationship between ei65 *murder* and *curious*, which it does, as well as e.g. ei65 *murder* and ei68 *wrote*, which is also a correct description of that temporal relationship. However, these relations are not in the gold standard annotation (despite being correct interpretations of the text) and so they present as false positives. Because manual examination of all the false positives to detect errors of this kind would be time consuming, the 26.2 % figure that comes from automatic evaluation must be seen as a lower bound.

For a more concrete evaluation, one can constrain the set of signal associations considered to that described by TLINKs in the document. That is, we assume that events and timexes are known, and also that interval pairs (as in TLINK arguments) have been identified, and that the remaining tasks in a document's TimeML annotation are signal annotation and then TLINK relation type assignment. To this end, one only considers pairs of intervals that are also found in the gold standard. Thus, the evaluation problem is constrained somewhat, excluding the implicit temporal relation identification stage the initial evaluation includes. Therefore, this is referred to as the "constrained joint approach". It is implemented by, instead of using a window to choose interval pairings for consideration, using the pairing suggested in each of the annotated TLINKs.

In this case, there are 136 gold standard entities again. Result are given in Table 5.31. The system finds 99 signalled interval pairs that have arguments corresponding to a TLINK in the gold standard. Of these 99, 88.9 % (88) are correct annotations (e.g. precision is 88.9 %); the remaining 11 are spurious. This gives a recall of 64.7 % and F1 of 74.9 %. We describe these with F1 and not the Matthews correlation coefficient often associated with evaluating binary classifiers because the set of true negatives is very large in this case but not very interesting, and F1 does not take them into account.

In summary, using no signal information from the gold standard and simply relying on models for signal annotation, we achieve a 74.9 % F1 rate for the overall joint task of identifying temporal signal expressions and linking each expression found to a pair of intervals that it temporally co-ordinates.

Table 5.31 Details of the constrained joint approach to signal annotation

Signal/TLINK associations	Count	Proportion (%)
In N45	136	–
Found	99	–
Found, both args in N45	88	88.9
Signal in N45, new TLINK assoc	0	0.00
Found based on new signals	11	11.1

5.8.2 Combined Signal Annotation and Relation Typing

We know that signals are helpful in informing TLINK labelling. We also know that we can automatically annotate signals, to a reasonable degree of accuracy. It remains to be seen whether this degree of accuracy is sufficient for automatically-created signal annotations that are of overall help in TLINK labelling. It may be that the TLINK labelling information provided by signals is offset by imperfect automatic signal annotation, or that false positives in signal annotation provide misleading and counter-productive information to TLINK labelling.

In this section, experiments are reported whose aim is to determine whether automatic signal annotation has an impact on the overall task of TLINK labelling. We take the N45 section of the AQUAINT corpus as the dataset. It is curated to add missing signals, intervals and associations (details in Table 5.32). Two experiments are conducted. The first, a baseline, is over the manually signal-augmented version of the N45 docs (AQN45-sig) using a link labelling model trained on TB-sig, including no signal-specific features. This ignores temporal signals and represents the situation where a gold standard annotation is performed and a model learned without any signal information, and evaluated over unseen data. The second experiment uses TB-sig to learn models for signalled and non-signalled TLINKs, using the signal features described in Sect. 5.3.1, and then evaluates the performance of these models at labelling their respective parts of the automatically signal annotated version of N45 described in Sect. 5.8.1. This represents the scenario of having already annotated events, timexes and pairing intervals, then doing automatic signal annotation on unseen data, and evaluates how helpful these signal annotations are for TLINK labelling. We exclude new TLINKs identified in the course of automatic signal association, as we have no gold standard the relation type of these. The version of N45 with automatically generated signal annotations is referred to as AQN45-auto.

The distribution of interval pair types and TLINKs in the training data, TB-sig, is shown in Table 5.33. Similar data for evaluation corpora is in Table 5.32.

Table 5.32 TLINK stats over corpora used for extrinsic evaluation

Corpus	TLINKs	Non-signalled	Signalled	Signal %
AQN45	1 048	932	116	11.1 %
AQN45-sig	1 062	915	147	13.8 %

Table 5.33 Training dataset sizes from TB-sig used for signal annotation models

Interval types	Non-signalled	Signalled
Event-event	3 179	343
Event-time	2 299	529
Time-time	126	14

Table 5.34 TLINK labelling accuracy over corpora used for extrinsic evaluation. The baseline is the overall most-common-class for TLINKs in the training data (TB-sig). Interval text features are not included. There were no timex-timex links. The difference between the first two rows shows the impact that this total asignal discrimination and association approach has on TLINK labelling accuracy

Corpus	Subset of links	Event-Event (%)	Event-Time (%)	Overall (%)	Baseline (%)
AQUAINT N45 plain	All	44.0	56.4	55.8	28.9
AQN45-auto	All	62.0	58.4	58.6	28.9
AQN45-auto	Unsignalled	50.0	58.6	58.5	28.4
AQN45-auto	Only signalled	66.7	56.8	59.2	32.0
AQN45-sig	Only signalled	70.5	72.2	71.64	32.8

It can be seen that TLINKing based on automatic signal annotations, detailed in the second row (AQN45-auto/all) of Table 5.34, performs better than TLINKing with no signal information (the first row). The approach is therefore effective.

However, signalled TLINKs in the gold standard are still labelled substantially better than when automatic signal annotations are used (compare the fourth and fifth rows). Event-event links tend to draw particular benefit from signal annotations (see second and third columns), and this is still the case with automatic signal annotations; 66.7 % accuracy was achieved on the signalled event-event links, and 70.5 % using gold-standard links, compared to only 44.0 % labelling accuracy without any signal information. Overall, event-event temporal relation typing performance on this dataset increased from 44.0 % accuracy ignoring signals to 62.0 % when using automatically annotated signals – an 18.0 % performance increase, or 32.1 % error reduction.

The N45 part of the AQUAINT corpus unfortunately has a much lower event-event: event-timex TLINK ratio than TimeBank, with only 50 event-event versus 1 012 event-time links (4.71 % of the whole). For comparison, TB-sig has 2 828 event-time links to 3 522 event-event; event-event comprise 55.5 % of links. The bias in N45 has therefore led to an underestimate of the extra impact that signal information has on general event-event labelling. Nonetheless, the results confirm the efficiacy of the automatic signal extraction method, and show an overall 2.8 % absolute improvement in TLINK labelling over data without signals.

5.9 Chapter Summary

Temporal signals are an important source of information for temporal relations.

This chapter presented a principled investigation into temporal signals and the role they play in relating and ordering events and times within discourse.

It first presented a linguistic account for temporal signals, followed by a demonstration of their utility in the relation typing task, with a prototype supervised learning approach to temporal relation typing with signals that achieved error reduction of 53 % compared to the same system without signal information.

Given this strong motivation for exploring signals, a corpus analysis of temporal signals was conducted, examining an existing TimeML-annotated corpus. This was followed by a brief attempt at automatic temporal signal annotation which quickly revealed insufficient quality in signal annotations. As a result, the corpus was re-annotated with extra signals, including the events, timexes and temporal relations that the new signals required. This resource is made publicly available, as TB-sig.

Having a strong corpus, an approach for automatic signal annotation could be developed. This was taken as a two-part task. Firstly, as many signal expressions are polysemous, one must determine which occurrences of candidate signal words occur having a temporal sense. This was achieved with 83.0 % precision. Secondly, given a signal, one must determine which temporal intervals it co-ordinates. Two approaches to this problem were addressed – one considering intervals one at a time and ranking them, then assuming that the top two are linked, and another considering each possible pair of intervals. The interval pair approach worked best, achieving 83.5 % precision.

Having developed both stages of the signal annotation mechanism, these were evaluated jointly against a new gold-standard signal corpus derived from the AQUAINT TimeML corpus. With the least-constrained, hardest evaluation technique, 64.7 % of the gold-standard annotations were found automatically by the discrimination/association system proposed in this chapter.

Finally, with a full signal annotation system developed, the impact of automatic signal annotation on the overall task of temporal relation typing was evaluated. Results were positive. Adding automatic signal annotations and then feature representations of these automatically-found signals improved the absolute performance of a temporal relation type classifier by 18 % for event-event links and 2.0 % for event-time links.

In summary, we showed that temporal signals were useful in temporal relation typing, and developed approached for automatically annotating them, which performed well enough to give a net performance increase in the temporal relation typing task.

References

1. Hitzeman, J.: Semantic partition and the ambiguity of sentences containing temporal adverbials. Nat. Lang. Seman. **5**(2), 87–100 (1997)
2. Ho-Dac, L., Péry-Woodley, M.: Temporal adverbials and discourse segmentation revisited. In: Multidisciplinary Approaches to Discourse (2008)
3. Bestgen, Y., Vonk, W.: Temporal adverbials as segmentation markers in discourse comprehension. J. Mem. Lang. **42**(1), 74–87 (1999)
4. Brée, D., Smit, R.: Temporal relations. J. Seman. **5**(4), 345 (1986)

5. Brée, D., Feddag, A., Pratt, I.: Towards a formalization of the semantics of some temporal prepositions. Time Soc. **2**(2), 219 (1993)
6. Schlüter, N.: Temporal specification of the present perfect: a corpus-based study. Lang. Comput. **36**(1), 307–315 (2001)
7. Vlach, F.: Temporal adverbials, tenses and the perfect. Linguist. Philos. **16**(3), 231–283 (1993)
8. Hitzeman, J.: Text type and the position of a temporal adverbial within the sentence. In: Proceedings of the 2005 international conference on Annotating, extracting and reasoning about time and events, pp. 29–40. Springer (2005)
9. Derczynski, L., Gaizauskas, R.: A corpus-based study of temporal signals. In: Proceedings of the Corpus Linguistics conference (2011)
10. Setzer, A., Gaizauskas, R.: Annotating events and temporal information in newswire texts. In: Proceedings of the Second International Conference On Language Resources And Evaluation (LREC-2000), Athens, Greece, vol. 31 (2000)
11. Derczynski, L., Gaizauskas, R.: Analysing temporally annotated corpora with CAVaT. In: Proceedings of the Language Resources and Evaluation Conference, pp. 398–404 (2010)
12. Derczynski, L., Gaizauskas, R.: Using signals to improve automatic classification of temporal relations. In: Proceedings of the ESSLLI StuS (2010)
13. Derczynski, L., Gaizauskas, R.: Temporal signals help label temporal relations. In: Proceedings of the annual meeting of the Association for Computational Linguistics. Association for Computational Linguistics (2013)
14. Mani, I., Verhagen, M., Wellner, B., Lee, C., Pustejovsky, J.: Machine learning of temporal relations. In: Proceedings of the 21st International Conference on Computational Linguistics and the 44th annual meeting of the Association for Computational Linguistics, p. 760. Association for Computational Linguistics (2006)
15. Bethard, S., Martin, J., Klingenstein, S.: Timelines from text: identification of syntactic temporal relations. In: Proceedings of the International Conference on Semantic Computing, pp. 11–18 (2007)
16. Zipf, G.: The Psycho-biology of Language. Houghton-Mifflin, Boston (1935)
17. Quirk, R., Greenbaum, S., Leech, G., Svartvik, J., Crystal, D.: A Comprehensive Grammar of the English Language, vol. 1. Longman, New York (1985)
18. Dorr, B., Gaasterland, T.: Summarization-inspired temporal-relation extraction: tense-pair templates and treebank-3 analysis. Technical Report. CS-TR-4844, University of Maryland, College Park, MD, USA (2006)
19. Mani, I., Hitzeman, J., Richer, J., Harris, D., Quimby, R., Wellner, B.: SpatialML: annotation scheme, corpora, and tools. In: Proceedings of LREC, vol. 8 (2008)
20. Verhagen, M., Saurí, R., Caselli, T., Pustejovsky, J.: SemEval-2010 task 13: TempEval-2. In: Proceedings of the 5th International Workshop on Semantic Evaluation, pp. 57–62. Association for Computational Linguistics (2010)
21. Reichenbach, H.: The tenses of verbs. Elements of Symbolic Logic. Dover Publications, New York (1947)
22. Stevenson, M., Wilks, Y.: Word sense disambiguation. The Oxford Handbook of Computational Linguistics, pp. 249–265. Oxford University Press, Oxford (2005)
23. Navigli, R.: Word sense disambiguation: a survey. ACM Comput. Surv. **41**(2), 1–69 (2009)
24. Lapata, M., Lascarides, A.: Learning sentence-internal temporal relations. J. Artif. Intell. Res. **27**(1), 85–117 (2006)
25. Charniak, E.: A maximum-entropy-inspired parser. In: Proceedings of the 1st North American chapter of the Association for Computational Linguistics conference, pp. 132–139. Morgan Kaufmann Publishers Inc. (2000)
26. Kiss, T., Strunk, J.: Unsupervised multilingual sentence boundary detection. Comput. Linguist. **32**(4), 485–525 (2006)
27. Klein, D., Manning, C.: Fast exact inference with a factored model for natural language parsing. Adv. Neural Inf. Process. Syst. **15**, 3–10 (2003)
28. Marcus, M., Kim, G., Marcinkiewicz, M., MacIntyre, R., Bies, A., Ferguson, M., Katz, K., Schasberger, B.: The Penn Treebank: annotating predicate argument structure. In: Proceedings

of the workshop on Human Language Technology, pp. 114–119. Association for Computational Linguistics (1994)
29. Bies, A., Ferguson, M., Katz, K., MacIntyre, R., Tredinnick, V., Kim, G., Marcinkiewicz, M., Schasberger, B.: Bracketing guidelines for Treebank II style Penn Treebank project. University of Pennsylvania (1995)
30. Blaheta, D., Charniak, E.: Assigning function tags to parsed text. In: Proceedings of the 1st North American chapter of the Association for Computational Linguistics conference, p. 240. Morgan Kaufmann Publishers Inc. (2000)
31. Musillo, G., Merlo, P.: Assigning function labels to unparsed text. In: Proceedings of RANLP'05 (2005)
32. Blaheta, D.: Function tagging. Ph.D. thesis, Department of Computer Science, Brown University (2004)
33. Gabbard, R., Marcus, M., Kulick, S.: Fully parsing the Penn Treebank. In: Proceedings of the main conference on Human Language Technology Conference of the North American Chapter of the Association of Computational Linguistics, pp. 184–191. Association for Computational Linguistics (2006)
34. Lintean, M., Rus, V.: Naive bayes and decision trees for function tagging. In: FLAIRS Conference, pp. 604–609 (2007)
35. Marcus, M., Marcinkiewicz, M., Santorini, B.: Building a large annotated corpus of English: The Penn Treebank. Comput. Linguist. **19**(2), 330 (1993)
36. Quinlan, J.: C4. 5: Programs for Machine Learning. Morgan Kaufmann, San Mateo (1993)
37. Freund, Y., Schapire, R.: A decision-theoretic generalization of on-line learning and an application to boosting. J. Comput. Syst. Sci. **55**(1), 119–139 (1997)
38. Freund, Y., Schapire, R.: Experiments with a new boosting algorithm. In: Machine Learning: Proceedings of the Thirteenth International Conference, pp. 148–156 (1996)
39. De Marneffe, M., MacCartney, B., Manning, C.: Generating typed dependency parses from phrase structure parses. In: Proceedings of the International Conference on Language Resources and Evaluation (2006)
40. Verhagen, M.: Times Between The Lines. Ph.D. thesis, Brandeis University (2004)
41. Kanhabua, N., Nørvåg, K.: Improving temporal language models for determining time of non-timestamped documents. In: Research and Advanced Technology for Digital Libraries, pp. 358–370. Springer (2008)

Chapter 6
Using a Framework of Tense and Aspect

> For years I have endeavored to break through the veil which shrouded it, and at last the time came when I seized my thread and followed it.
>
> *The Final Problem*
> SIR ARTHUR CONAN DOYLE

6.1 Introduction

This chapter investigates a linguistic framework for tense and aspect. Analysis of the temporal relation typing problem in Chap. 4 suggested two directions for investigation. Temporal signals were one of these; tense and aspectual differences were the second prevalent category. Having investigated temporal signals in Chap. 5, this chapter is dedicated to the other major source of temporal ordering information in difficult links.

Tense and aspect are used to describe temporal aspects of events which are expressed with verbs. It is intuitive that tense and aspect will be of some value for determining the type of temporal relation that holds between two verb events, and evidence in human-annotated corpora supports this intuition.

Event-event relations are the hardest to label (Chap. 4). Around 45 % of links in TempEval (a temporal annotation evaluation exercise, see Sect. 3.4.4.4) event-event tasks cannot reliably be labelled automatically (see Sect. 4.2.2). Further, verb-verb links make up a significant amount of the difficult links identified in Sect. 4.2.

Relations involving at least one argument with tense or aspect information are prevalent. They are also difficult to label. Verb-verb links make up around a third of TimeBank's TLINKs, and tensed verb-verb links the largest share of that set, so of all verb-verb relations, the majority are between two tensed verbs.

Ordering time expressions and events in the same sentence is a also somewhat difficult task. In TimeBank, almost half of all TLINKs are between a time and event. Of these, half are between an event and timex in the same sentence, where the timex is a date or time.

L.R.A. Derczynski, *Automatically Ordering Events and Times in Text*,
Studies in Computational Intelligence 677, DOI 10.1007/978-3-319-47241-6_6

Table 6.1 Frequency of TimeML tense and aspect on verb events in TimeBank

Tense	Aspect	Count
PAST	NONE	1975
PRESENT	NONE	803
INFINITIVE	NONE	762
PRESPART	NONE	360
PRESENT	PERFECTIVE	270
FUTURE	NONE	262
PRESENT	PROGRESSIVE	162
PASTPART	NONE	150
PAST	PERFECTIVE	88
NONE	PERFECTIVE	20
PAST	PROGRESSIVE	19
PRESENT	PERFECTIVE_PROGRESSIVE	17
FUTURE	PROGRESSIVE	5
FUTURE	PERFECTIVE	4
NONE	PROGRESSIVE	3
NONE	PERFECTIVE_PROGRESSIVE	2
PASTPART	PERFECTIVE	2
PAST	PERFECTIVE_PROGRESSIVE	1
PRESPART	PERFECTIVE	1

Data-driven approaches to the relation typing task are hampered in two ways. Firstly, there is a shortage of ground truth training data, which is in turn partially due to the high cost of annotation. As [1] point out, this leads to low volumes of instances for many combinations of tense and aspect values for pairs of events (see Table 6.1), potentially hampering automatic hypothesis learning. Secondly, the variation of expression annotatable using TimeML is relatively limited, describing three "tenses"[1] (past and past participle, present and present participle, and future) and three "aspects" (none, perfective and progressive). This markup language may be insufficiently descriptive to capture the relations implied by all the variations in linguistic use of tense and aspect.

Reichenbach [2] offers a theoretical framework for analysis of tense and aspect that can be used to predict constraints on temporal orderings between verb events based on their tense and aspect, and also between times and tensed verbs. Applying Reichenbach's framework requires tense and aspect information, which is provided in TimeML (meaning that it might be possible to apply this framework without a major annotation effort).

[1]In TimeML v1.2, the tense attribute of events has values that are conflated with verb form. This conflation is deprecated in versions of TimeML more recent than that in which TimeBank is annotated.

Application of the framework gives a partial idea of the temporal ordering between a suitable pair of events or an event and timex (except durations and sets). These rough orderings can be used to constrain of the set of possible TimeML relation types for any given pair. For example, a suggestion of "overlap" constrains possible TimeML relations to "simultaneous/includes/included_by".

It may be the case that machine learning methods are unable to make effective use of the tense information available in TimeBank. Phenomena such as tense shifts between events have been shown to help humans temporal ordering [3], and therefore may convey some temporal information. However, the percentage of links with tense shifts is roughly the same in the general case (40 % in TimeBank) and the difficult link set (36 %). As these figures are roughly the same, it may be that supervised approaches fail to make generalisations that take advantage of the information given in tense shifts.

Prior work has gone some way to determining the utility of tense in the relation typing task. The USFD system in TempEval-2007 [4] found that the supplied tense was not a helpful feature for event-timex linking (though aspect was), though that it did provide some benefit to event-event ordering when the events were in the same or adjacent sentences.

Reichenbach's framework may offer a method for determining or approximating temporal orderings over this significant part of the difficult link set (and also in the general case). In this chapter, we offer a full account of Reichenbach's framework in the context of TimeML, and investigate how consistent the framework is with gold-standard temporally annotated data, before offering methods for integrating it into a temporal relation typing approach.

The rest of this chapter is structured as follows. Firstly, we discuss in abstract terms a conceptual model for time. Second, there is an introduction to Reichenbach's framework and a description of how it interacts with temporal expressions as well as verb events, followed by a summary of related work. Next, validation of the framework is attempted by describing how the framework can be related to TimeML and then an evaluation of it against ground truth temporal relation type information. The framework's relation type constraints are then applied to the temporal relation typing task alongside data from TimeML annotations, as part of a machine learning approach to relation typing, and results presented. It is found that Reichenbach framework is potentially helpful. To allow inclusion of what the framework provides that is not in TimeML already, an annotation scheme for the framework is introduced (RTMML) which may also be used as an extension to TimeML. Finally, the chapter concludes with a discussion of applications of the framework and future work.

6.2 Timelines in Language

Time, as experienced and expressed by humans, seems to be linear. Events begin and end at points along this line, through which travel is always unidirectional; each event's end can come no earlier than its beginning.

Time is often described using the same language as space, as touched upon in Sect. 5.4.4. We talk about *time travel*, use words such as *faster*, *before* and *at* and specify directions such as *forward* and *backward*. The linguistic relation between expression of time and space is sometimes taken to extremes; some have suggested that we travel through time facing backwards, because we can only see the past and not the future [5]. The spatio/temporal polysemy is even learned by classifier models when attempting to detect temporal usages of words (Sect. 5.6.4.3). This linguistic similarity is rooted in the way that humans understand non-literal motion (such as in temporal transitions) using the same cognitive resources as we understand literal (e.g. spatial) motion [6].

Given that time is a linear and effectively continuous [7] dimension which progresses unidirectionally [8] but can be conceived of in either direction [9], we talk about its description in language with a model of time as uni-dimensional (cf. McTaggart's A-series [10]).

As a line is a conceptually simple spatial representation of a single linear dimension (such as time), we shall describe our temporal dimension by means of a "**timeline**". We are constantly at a point that we refer to as the present. This point exists on the timeline as a separator between the past and the future. Our timeline can thus be described as three non-overlapping parts: past, present and future.

The time at which an utterance is heard or read is always the present. Some way is required of referring to events at points on a timeline that happen any time but the perceiver's present. One can perhaps define a method of absolute description of positions on a timeline, maybe by use of a calendar or clock[2] to determine origin locations. However, the attachment to every event of a label defined using an external scale causes event descriptions to be awkward both to write and to read (even ignoring the overhead of temporal scale creation, maintenance and reference). A potentially simpler mechanism is to describe events relative to each other; one may like to talk of things happening either at present, in the part of the timeline before it, or the part coming later.

These three parts correspond directly to the rudiments of tense in language; the past tense, present tense and future tense permit expression of events within the past, at the present, or within the future part of a timeline (cf. McTaggart's B-series). Thus, simple tense usage allows positioning of events within regions on a timeline relative to the present; and so, in that it describes temporally relative points, tense is inherently deictic [11, 12]. The tenses corresponding to these three categories are known as **absolute tenses**.

Given such a tense structure, one may identify two temporal points upon the timeline. One is the time at which the description of the event is uttered or perceived, and the other, that may be in any of the three timeline parts, corresponds to the time

[2]In fact, each of these "absolute references" eventually relies upon events. A year is the event of a full cycle of the earth around the sun, and a second is the duration of a certain number of caesium isotope decay events. The common era calendar is centred around an agreed point based on a described event; each day's start (e.g. midnight) is determined by the event of a specific angle of rotation of the earth upon its axis relative to the sun.

that the described action took place. This simple structure allows us to temporally express events relative to the present.

However, the ability to relate events to each other – critical to planning and story-telling – is still difficult with this system. If we are to mention an event and then express another event in terms of that (e.g. *The race will be over and I will have won*), one must be able to treat the first event as a sort of basis or origin for positioning the second. In this example, the *winning* happens in one of the three parts of a timeline where the "present" is at or after the race's completion. To express this, we need what amounts to double-deixis; there is one three-part structuring of the timeline where the present centres upon the time of utterance, and another with the present situated around the race's completion.

In language, this double-deixis can be accounted for in a system of tense and aspect. It is required not only to describe a primary event relative to its primary deixis, but also then to describe a secondary event relative to the primary event. This might involve a relocation of the listener such that the secondary event's temporal position is described in terms that they are familiar with – such as the 3-part past/present/future model – centred not upon the listener's present, but instead around the primary event described. In our example, the *winning* is described not relative to the time the sentence is uttered, but in terms of the event of the race's end.

As well as recognising divisions of past, present and future, we can describe this secondary structuring of a timeline around an event by use of anterior, simple and past tenses. These correspond to events described before, at or after the initially-described event. Continuing to use the race example, the race is over at some point in the future, and the *winning* happens before this – anterior to the primary event. As the primary event occurs in the future, we say that *I will have won* is in the *anterior future* tense. This gives us a tense system that allows the description both of events relative to now, and also of events relative to each other that is also readily describable using a timeline.

It is worth noting at this point that, being irrealis from the point of reference, the future tense is often considered a modality rather than a tense – certainly in English. This is echoed by McTaggart's argument for incoherence of the A-series (the absolute, external, ordered sequence of events) [10]; he essentially claims that time is incoherent, as we know that events have an innate ordering, so how could we not see what that ordering is? Any given event, as time advances, will be past and will have been future. Jaszczolt details with this another way, by putting forward that temporality is modal, with different tenses (or other representations of time) having varying degrees of certainty [13]. Both of these arguments hinge on the future being modal. In any event, one generally needs linguistic devices with which to describe the future, and tense is such a device, where the future is just one partition (the others being past and present).

6.3 Description of the Framework

The core of the framework comprises three abstract time points – speech time, event time and reference time – which are related to each other in terms of equality (e.g. simultaneity), precedence or succession. The tense and aspect of verbs are then described using these points, which we introduce properly next. Finally, interactions between verbs are formalised in terms of relations between the abstract time points of each verb. This section introduces the basic framework as proposed by Reichenbach, and then discusses its limitations and puts forward additional proposals for extending the framework.

6.3.1 Time Points

To describe a tense, Reichenbach introduces three abstract time points. Firstly, there is the speech time,[3] S. This represents the point at which the verb is uttered or written. Secondly, event time E is the time that the event introduced by the verb occurs. Thirdly, there is reference time R; this is an abstract point, from which events are viewed. Klein [15] describes it as “the time to which a claim is constrained”.

In Example 18, speech time S is when the author created the discourse (or perhaps when the reader interpreted it).

Example 18 By then, she had left the building.

Reference time R is *then* – an abstract point, before speech time, but after the event time E, which is the leaving of the building. In this sentence, one views events from a point in time later than they occurred. Therefore, the final configuration is $E < R < S$.

6.3.2 Reichenbachian Tenses

Reichenbach details nine tenses (see Table 6.2). The tenses detailed by Reichenbach are past, present or future, and may take a simple, anterior or posterior form. In English, these apply to single non-infinitive verbs and to verbal groups consisting of head verb and auxiliaries. The tense system describes abstract time points for each tensed verb and how they may interact, both for a single verb and with other events described by verbs.

In Reichenbach's view, different tenses specify different relations between E, R and S. Table 6.2 shows the six tenses conventionally distinguished in English. As

[3]For this book, speech time is equivalent to DCT, unless otherwise explicitly positioned by discourse. Under Fillmore's description [14], this is the same as always setting speech time S equal to encoding time ET and not decoding time DT.

Table 6.2 Reichenbach's tenses; from [16]

Relation	Reichenbach's tense name	English tense name	Example
$E<R<S$	Anterior past	Past perfect	I had slept
$E = R<S$	Simple past	Simple past	I slept
$R < E < S$ $R < S = E$ $R < S < E$	Posterior past		I expected that I would sleep
$E<S = R$	Anterior present	Present perfect	I have slept
$S = R = E$	Simple present	Simple present	I sleep
$S = R<E$	Posterior present	Simple future	I will sleep (Je vais dormir)
$S < E < R$ $S = E < R$ $E < S < R$	Anterior future	Future perfect	I will have slept
$S<R = E$	Simple future	Simple future	I will sleep (Je dormirai)
$S<R<E$	Posterior future		I shall be going to sleep

there are more than six possible ordering arrangements of S, E and R, some English tenses might suggest more than one arrangement. Reichenbach's named tenses names also suffer from this ambiguity when converted to $S/E/R$ structures, albeit to a lesser degree. When following Reichenbach's tense names, it is the case that for past tenses, R always occurs before S; in the future, R is always after S; and in the present, S and R are simultaneous. Further, "anterior" suggests E before R, "simple" that R and E are simultaneous, and "posterior" that E is after R. The flexibility of this framework is sufficient to allow it to account for a very wide set of tenses, including all those described by [17], and this is sufficient to account for the observed tenses in many languages. Past, present and future tenses imply $R < S$, $R = S$ and $S < R$ respectively. Anterior, simple and posterior tenses imply $E < R$, $E = R$ and $R < E$ respectively.

6.3.3 Verb Interactions

While each tensed verb involves a speech, event and reference time, multiple verbs may share one or more of these points. For example, all narrative in a news article usually has the same speech time (that of document creation). Further, two events linked by a temporal conjunction (e.g. *after* - see Chap. 5) are very likely to share the same reference time. Basic methods of linking between verb events or linking verbs to fixed points on a time scale are described below.

6.3.3.1 Special Properties of the Reference Point

The reference point R has two special uses. These relate to verbs in the same *temporal context* (see Sect. 6.3.4 below) and to the effect of time expressions on verbs.

Permanence

Firstly, when sentences are combined to form a compound sentence, tensed mean verbs interact, and implicit grammatical rules require tenses to be adjusted. These rules operate in such a way that the reference point is the same in all cases in the sequence. Reichenbach names this principle **permanence of the reference point**; *"We can interpret these rules as the principle that, although the events referred to in the clauses may occupy different time points, the reference point should be the same for all clauses"*. Figure 6.1 contains an example of this principle.

Positional

Secondly, when temporal expressions (such as a TimeML TIMEX3 of type DATE, but not DURATION) occur in the same clause as a verbal event, the temporal expression does not (as one might expect) specify event time E, but instead is used to position reference time R. This principle is named **positional use of the reference point**.

In Example 19, an explicit time (*10 o'clock*) determines our reference point through positional use.

Example 19 It was 10 o'clock, and Sarah had brushed her teeth.

The verb group *had brushed* is anterior past tense; that is, $E < R < S$. The event is complete before the reference time – that is, at any point until *10 o'clock* – and so the relation between the event and timex can be determined (*brushed* BEFORE *10 o'clock*).

6.3.3.2 Example Reichenbachian Verb-Verb Links

All three points from Reichenbach's framework are sometimes necessary to position an event on a timeline or in relation to another event. For example, they can help determine the nature of a temporal relation, or a calendar reference for a time. We illustrate this two brief examples.

*Example 20 In February 1917, the Germans landed their offensive. By April 26*th*, it was all over.*

Example 20 shows a temporal expression describing a day – April 26th. The expression is ambiguous because we cannot position it absolutely without knowing

"John **told** me the news" : "told" is simple past, so:
1. E = R < S

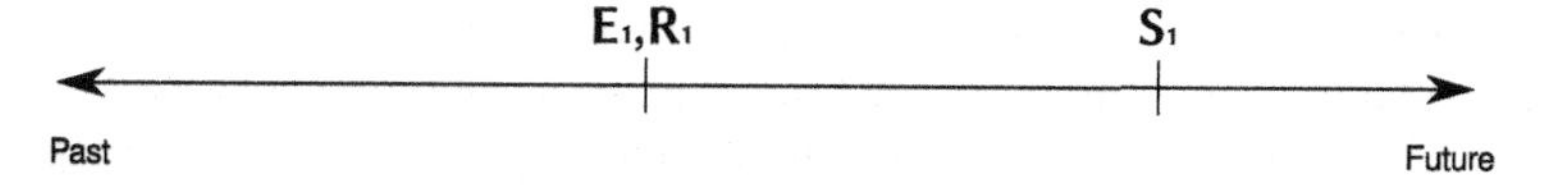

"I **had already sent** the letter" : "had already sent" is anterior past, so:
2. E < R < S

Both utterances have the same speech time.

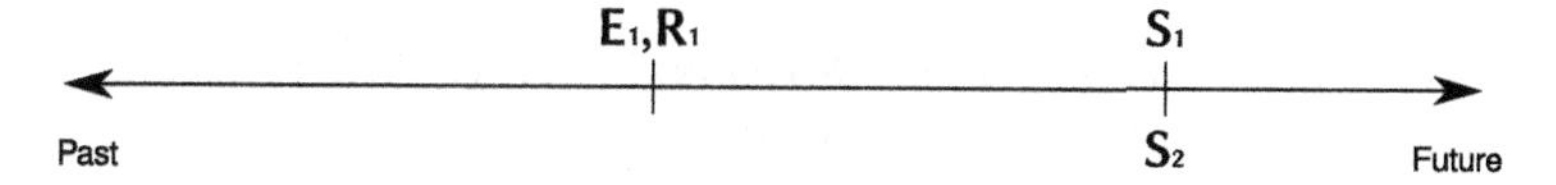

Because they are in the same clause, by permanence of the reference point, reference time is also shared.

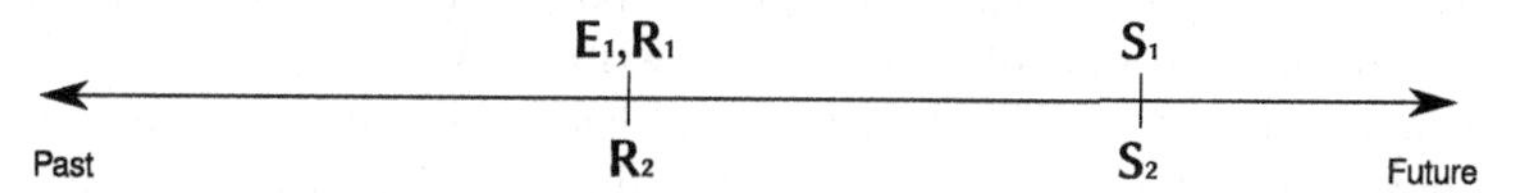

We know that $E_2 < R_2$.

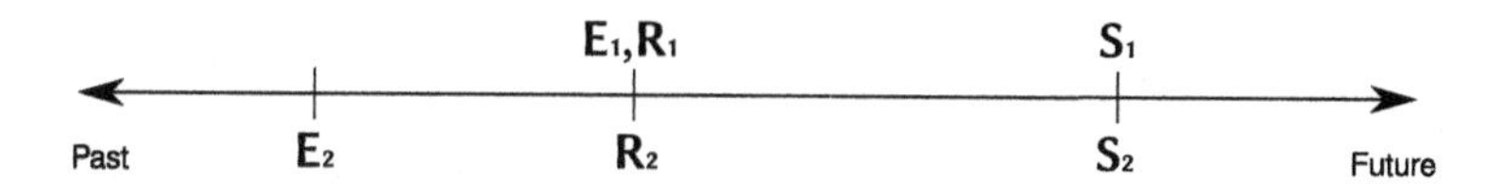

Therefore, using Reichenbach's framework and simple reasoning, we can determine that E_1 happens after E_2 from the tenses and context of these events.

Fig. 6.1 An example of permanence of the reference point

which year it refers to. This type of temporal expression is interpreted with respect to reference time, not with respect to speech time [18]. Without a time frame for the sentence (presumably provided earlier in the discourse), we cannot determine which year the date is in. If we are able to set bounds for R in this case, the time in Example 20 will be the April 26th adjacent to or contained in R; as the word *by* is used, we know that the time is the April 26th following R, and can normalise the temporal expression, associating it with a time on an absolute scale.

Example 21 John <u>told</u> me the news, but I had already <u>sent</u> the letter.

Example 21 and Fig. 6.1 show a sentence with two verb events – *told* and *had sent*. Using Reichenbach's framework, these share their speech time S (the time of the sentence's creation) and reference time R, but have different event times. In the first verb, reference and event time have the same position. In the second, viewed from when John told the news, the letter sending had already happened – that is, event time is before reference time. As reference time R is the same throughout the sentence, we know that the letter was sent before John mentioned the news. Describing S, E and R for verbs in a discourse and linking these points with each other (and with times) is the only way to ensure correct normalisation of all anaphoric and deictic temporal expressions, as well as enabling high-accuracy labelling of some temporal links.

Example 22 contains a more advanced example. It shows a pair of temporally related verbs taken from the list of difficult links found earlier (see Sect. 4.3.1).

Example 22 A committee of outside directors for the Garden City, N.Y., unit is evaluating$_{e1}$ *the proposal ; the parent asked*$_{e2}$ *it to respond by Oct. 31.*

One can determine the temporal relation between events `e1` and `e2` from the tenses in this sentence without particularly complex reasoning. In the example, `e1` is present progressive, and `e2` is past tense. Te end point of *evaluating* (`e1`) is after the end of `e2` and after the time of the example's writing. We can also see that the end of `e2` is in the past – the *asked* started and finished before document creation time (DCT), and certainly finished before *evaluating* finishes. This tense-based reasoning gives a constrained set of temporal relation types.

6.3.4 Temporal Context

In the linear order that events and times are introduced in discourse, speech and reference points persist until changed by a new event or time. Observations during the course of this work suggest that the reference time from one sentence will roll over to the next sentence, until it is repositioned explicitly by a tensed verb or time. To make discussion of sets of verbs with common reference times easy, we call each of these groups a **temporal context**.

To cater for subordinate clauses in cases such as reported speech, we add a caveat – S and R persist as a discourse is read in textual order, for each temporal context. A context is an environment in which events occur, and may be the main body of the document, a tract of reported speech, or the conditional world of an *if* clause [19]. For example:

Example 23 Emmanuel had said "This will explode!", but changed his mind.

Here, *said* and *changed* share speech and reference points. Emmanuel's statement occurs in a separate context, which the opening quote instantiates, ended by the closing quote (unless we continue his reported speech later), and begins with an S that occurs at the same time as *said* – or, to be precise, *said*'s event time E_{said}.

Temporal contexts may be observed frequently in natural language discourse. For example, the main body of a typical news article shares the same reference point, reporting other events and speech as excursions from this context. Each conditional world of events invoked by an "if" statement will share the same context. Events or times linked with a temporal signal will share a reference point, and thus be explicitly placed into the same temporal context.

As described in Chap. 4 of [19] in his description of the sequence of tenses with regard to Reichenbach's framework, permanence of the reference point does not apply between main events and embedded phrases, relative clauses or quoted speech. These occur within a separate temporal context, and it is likely that they will have their own reference time (and possibly even speech time, for example, in the case of quoted speech). In order to apply permanence of the reference point, it ought only be applied within the same temporal context. Verbs to which permanence may be applied are said by Reichenbach to be those to which the grammatical rules of the **sequence of tenses** (an abstract set of grammatical rules not described in his paper) apply. Different contexts will have a consistent reference point, and so permanence of the reference point may be applied to verbs within that context in order to gain information about their temporal relations. Permanence does not apply across different temporal contexts.

Dowty [20] hints at the concept of temporal context with the idea of the **temporal discourse interpretation principle** (TDIP). This states:

> Given a sequence of sentences $S_1, S_2, \ldots, S_n$ to be interpreted as a narrative discourse, the reference time of each sentence S_i (for i such that $1 < i - n$) is interpreted to be:
>
> (a) a time consistent with the definite time adverbials in S_i, if there are any;
>
> (b) otherwise, a time which immediately follows the reference time of the previous sentence S_{i-1}.

The TDIP accounts for a set of sentences which share a reference and speech point. However, as with other definitions of temporal context, this principle involves components that are difficult to automatically determine (e.g. "consistent with definite time adverbials"). Miller et al. [21] may offer a parallel account of temporal context, in their definition of narrative containers, though it is down to empirical comparison to answer this question.

As discussed above, Temporal context describes the events which may temporally linked using Reichenbach's framework in order to helpfully constrain the set of temporal relations between each pair. It is therefore useful to automatic relation typing approaches to know the bounds of each temporal context. However, this information is not present in TimeML annotations and not readily available from discourse. This gives the problem of having to model temporal context, in order to decide which event verb-event verb TLINKs to apply the framework.

Modeling temporal context requires the grouping of tensed verb event pairs so that only those in which both events are in the same temporal context are together. Simple techniques for achieving this could work on sentence proximity. In TimeBank, there are 1 167 event-event TLINKs where both arguments are tensed verbs, of which 600 are in the same sentence and a further 313 are in adjacent sentences. Further techniques for temporal context modelling are detailed in experiments below. Proximity alone may not be sufficient, given this chapter's earlier observations about quoted speech, re-positioning of the reference point and so on; however, it is a simple starting point.

While positional use of the reference point indicates a new (or change to an established) temporal context, and permanence of the reference point can only persist within the same temporal context, the principle of quoted speech (above) permits linking across some temporal contexts.

6.3.5 Quoted Speech

The framework can also be used to described adjustment of speech, reference and event time around reported, quoted speech. Although not mentioned in Reichenbach's original account, the principle emerges directly from his framework, and is as follows. When a verb is used to initiate quoted, reported speech, the speech time for that quote is equivalent to the event time of the initiating verb.

Example 24 shows two verb events: one initiates quoted speech (*told*), and the other is within this reported speech (*hold*).

Example 24 This morning General Powell told reporters, "We will hold a press conference shortly."

In this case, the event time of *told* corresponds to the speech time of *hold*. This form of reasoning allows us to connect events within quoted speech to those outside it. It may be referred to as **positional use of the speech point**. Just as with positional use of the reference point, where another entity determines how the *reference point* should be interpreted, positional use of the speech point occurs when another entity (in this case an event) determines how the *speech point* should be interpreted.

Exposition of the principle benefits from [19]'s modestly extended definition of speech time, as follows:

> The key to the analysis is the recognition that the S point has two related yet logically distinct properties: (i) it is a deictic anchor and (ii) it has a default interpretation in which it is mapped onto the utterance time if not otherwise interpreted.
>
> Distinguishing these two properties of the S point permits the formulation of a sequence of tense rule for embedded finite clauses. In this case, the rule associates an embedded point, S_{n-1}, with a higher point, E_n.

6.3.6 Limitations of the Framework

This section contains a discussion of some shortcomings of Reichenbach's tense framework and – where relevant – the proposed solutions.

6.3.6.1 Limited Tenses

The included tenses and aspects are insufficiently expressive to cover the gamut of linguistic expressions of temporality. One may look at lexical semantic models of tense and aspect in English to discover a wider inventory of possible tenses and aspects in that language [22], or examine other languages with richer aspect systems to see what the framework glosses over in those cases (e.g. [23]). Limitations of the Reichenbachian perfect can be seen from Table 6.2, where there is more than one triple that corresponds to the future perfect. Nevertheless, many tense and aspect systems can be described in terms of Reichenbach's framework, albeit not always as a 1:1 mapping.

6.3.6.2 Progressive Aspect

The progressive is used for events that have both a start and end and are currently ongoing; that is, in-progress activities. This makes it possible to refer to points within an event. However, Reichenbach's framework is point-based, and point-based temporal algebras generally assume that when point events are referenced, they are only referenced in terms of being before, after or simultaneous with another temporal entity. This makes it difficult to accurately represent more complex verbal event structures. Introducing interval reasoning to the framework can help (that is, dealing with intervals in terms of start and end points, instead of a single point for the whole), although it is sufficient to achieve this through treating events as a coupled start and end point (where the start is never after the end). This has the advantage of permitting semi-interval type reasoning (see Sect. 3.2.0.3). We discuss this further in Sect. 6.4.2

6.3.6.3 *On* Dates

Positional use of the reference point tells us that R is equivalent to a timex in the clause, if given. Because the algebra the framework uses to describe tenses is point-based, the start and end of the given time period are equal to the start and end of the reference time. This gives problems when a described event takes place during a provided timex, but does not have the same start and stop times. Example 25 is taken from [24]:

Example 25 Mary left England on May the 22nd, 1979

In this case, although Reichenbach's framework tells us that $R = E$ and that R is equivalent to *May the 22nd, 1979*, it is false that the leaving – E – took place simultaneously with the date; rather, it was a subpart of this 24 h interval. One solution to this unintuitive behaviour is to replace the reference point with a reference interval, having distinct start and end points if required.

6.3.6.4 Non-English Tense System

Some languages are difficult to accommodate in Reichenbach's framework. To accommodate Russian, for example, one must make specific and extensive additions to the framework, including binary temporal relations between points for each verb [25]. Such a system can be extended to cover a large range of Slavic languages [26], though is too complex to implement for a first attempt at automated temporal annotation using Reichenbach's framework.

Further, Reichenbach's framework is less useful given a language that has a limited tense system. It relies on a richness of expression placed in verb tenses. Without this richness, the value of applying the framework is reduced. For example, Chinese does not inflect verbs to express tense, but rather uses grammatical constructions, particles and temporal adverbials to describe time. The system is still somewhat less complex (regarding Reichenbach's framework) than that of English or French. The habitual, present, present progressive and stative can all be expressed the same way.

Example 26 我吃吗 (wǒ chī mǎ) – "I eat horse"

A simple sentence is given in Example 26. This can be interpreted in English as *"I prefer to eat horse"*, *"I am currently eating horse"* or *"I will eat horse"*, *"I ate horse"*; contextual markets are required for clarification. The default interpretation is that of simple present tense. Past tense can be signified with guō (过), and completion with le (了), both of are placed directly after the verb. It is therefore possible to capture the relation between speech and event points, and we can determine if the reference point is after the event or not. There is nothing to clarify the difference between simple and anterior tenses, and (as in English) the simple present is also used to indicate habitual truths (e.g. *I eat horse*). However, unlike English, the simple present progressive (e.g. *I am eating horse*) looks identical to the habitual use. Further information is expressed through temporal adverbials and not considered tense. The general lack of inflection or cohesive verb groups suggests that Reichenbach's framework can only be applied to Chinese in a limited fashion, decreasing its general utility.

6.3.6.5 Split Reference Point

Some tensed temporal descriptions of events are difficult to framework with just a single reference point. For example, from [27]:

Example 27

- *"I shall have been going to see John."* (that is, there is some point in the past at which I anticipated seeing John; note this is not a description of habitual behaviour)
- $S < R_1 < E < R_2$

It is true that the tenses and abstract points provided by the three-point framework are insufficient to capture this statement, without invoking an extra verb event.

However, in TimeBank no such contrived utterances were found during candid examinations or error analysis from applying the framework to predict TimeML relations.

6.3.6.6 Reification of the Reference Point

Tanaka [28] takes exception to the abstract nature of the reference point, and that it is never reified or explicitly lexicalised. He questions the requirement for reference time in a system of tense, and raises a few examples that are difficult to express using Reichenbach's framework. Tanaka's criticism and example are as follows.

Example 28

- Now Megumi will marry Kazuhiko next month.
- $S < E = R$

In Example 28, the temporal adverbial *next month* is used to position the reference point, R. With the tense used here – simple future – this also places E (the time of marrying) during *next month*, which is the correct interpretation. However, Tanaka suggests that the framework does not explain the influence of *Now* in this sentence; for which verbs does it fix the reference point? This criticism could be viewed as a variation on the requirement for two reference points to describe some verbs.

We can, in fact, provide a concrete solution in this case. One could attach *Now* to the auxiliary verb *will*, which provides a correct arrangement of points under Reichenbach's framework and is also an effective way of representing the situation in TimeML. It is not proposed that this is a satisfactory solution in terms of linguistic theory, rather, that it is a solution in computational for the purpose of automatically determining the nature of a given temporal relation.

6.4 Validating the Framework Against TimeBank

Having described Reichenbach's framework of tense and aspect and introduced related linguistic and temporal concepts, we now investigate how the framework compares with real data. Before applying Reichenbach's framework to the TimeML relation typing task, it is important to check if it is descriptively adequate. As it is possible to identify a set of candidate links where the argument types are of the right type (tensed verb events), the relation types of these can be compared with those suggested by the framework.

In order to evaluate its suggestions, temporal relation types suggested by the framework can be compared with a human-annotated ground truth, such as TimeBank. The framework can be applied to TLINKs where both arguments are tensed verbs, given tense and aspect information. This fits the difficult case identified in Chap. 4, that of event-event links involving some shift of tense. When ordering events based on positional use or permanence of the reference point, the set of TLINKs is further constrained to those where both arguments are in the same temporal context.

To compare the framework with TimeML-annotated resources, a number of decisions must be taken as part of an interpretation of the framework. Firstly, the Reichenbachian tense and aspect attributes do not directly match those in TimeML; some kind of mapping needs to be created between these two tense/aspect systems. One must convert a tense from TimeML into an arrangement of speech, event and reference point. Reichenbach suggests nine "basic" tenses and his system allows many arrangements of these points; TimeML separates tense and aspect and allows for values quite different to those included in Reichenbach's framework.

Secondly, Reichenbach is vague about temporal context. It is unclear from TimeML annotations alone which sets of verbs can be considered to be in the same "temporal context" (see Sect. 6.3.4). Reichenbach simply states that the framework is intended to follow the sequence of verbs. The descriptions of the "sequence of tenses" suggest it is difficult to implement programatically with current technology (see e.g. Chap. 4 of [19]), and require accurate identification of reported speech, embedded phrases, relative clauses, reference-time shifting temporal adverbials and so on. This presents a number of complex syntactic and linguistic scoping tasks that may be difficult to perform automatically. Therefore, one needs an approximation of temporal context in order to choose which verb pairs to attempt to relate.

Aside from these two decisions which help determine which event pairs to link and how to represent them, it is useful to construct a table describing temporal relation constraint according to the framework. The suggested type of relation between two events (or an event and a timex) – given their tense and aspect in Reichenbach's framework, and that permanence of the reference points holds between them – is not provided elsewhere, and some kind of relation matrix needs to be determined. To use tense and aspect values for temporal relation typing within the framework, we are concerned with possible arrangements of two event times given two verbs that represent these events, and need to describe the relation between event times. This provides a means to extract useful ordering information even in the situation that reference times do not match perfectly.

In the two-event sentence of Example 29, *fished* is anterior present with arrangement $E < S_1 = R_1$ and *eat* is simple future, with arrangement $S < R_2 = E_2$.

Example 29 *"I have fished$_1$; John will eat$_2$."*

The event times are located such that *fished* wholly precedes *eat* with relation to the speech time, regardless of reference time's situation, leading to the equivalent of a TimeML BEFORE relation. It is not always possible to suggest a relation, perhaps due to a lack of information; for example, two events in the simple past cannot be temporally ordered relative to one another without further information (e.g. in "I went to school, you went to church").

Note that *eat*$_2$ could be interpreted as Reichenbachian posterior present, with arrangement $S = R_2 < E_2$. This gives the same temporal ordering of events, but through transitivity permits a shared reference point (i.e. $R_1 = R_2$). In this situation, as is sometimes the case in English, it is not possible to decide precisely which of posterior present and simple future applies. However, this is of little impact in this toy example when we are concerned primarily with determining relations between events; the reference point is only a means to that end.

To record relation types ready for later look-up, a two-dimensional matrix is constructed, with each axis labelled using all possible combinations of tense and aspect values under whatever scheme the first decision's outcome permits. Each cell in this matrix contains the temporal relation between event times suggested by the tenses and aspects of its axes.

The rule of permanence of the reference point could potentially be applied to a large number of temporal relations (e.g. those where both arguments are verb events), and if helpful, is the rule that could have the highest impact. For this reason, we only examine relations between two events where both events are verbs that have some tense information.

Below are details of a minimal interpretation and also an advanced interpretation of the framework, including quantitative assessment of their agreement with TimeBank's event annotations.

6.4.1 Minimal Interpretation of Reichenbach's Framework

The only criterion for permanence rule applicability not present in TimeML annotation is whether or not a pair of events are in the same temporal context. This was approximated by only considering event-event links where both events were in the same or adjacent sentences. In TimeML, event-event links between events inside or outside quotes and conditional/intentional constructs are annotated using other mechanisms, such as the SLINK, and not included in the relation typing task addressed. A selection of 211 links from TimeBank that match this approximation to temporal context were then manually examined to see if temporal context actually applied. Of this 211, a majority (146 – 69.2 %) had both arguments in the same context.

These cases were identified manually as follows. Firstly, the search space was narrowed to verb-verb events within the same or adjacent sentences. A random sample of these was drawn for manual examination. Instances where one event lay in a different temporal context were then excluded. A shift in reference time for the events means that they are not in the same context, and this was generally caused by a timex, one event being in an embedded phrase or relative clause, a special sense of a verb (such as habitual or stative), or one argument being in reported speech that the other is not.

To the 146 manually-annotated same-context temporal relations, temporal relation constraints derived from Reichenbach's framework were applied, to see if the gold standard annotated TimeML relation was consistent with the suggested constraints.

Reichenbach's framework can return some temporal ordering information for event pairs given a pair of tensed verb arguments in the same temporal context. As the only relations available are precedence and equality (simultaneity), the possible return values are: BEFORE, AFTER, OVERLAPPING (which subsumes simultaneous) and VAGUE. The relation VAGUE is assigned when, for example, both events occur before reference time but nothing else is known; this is not enough to describe any kind of order between events. These values are coarser than the TimeML relation types, and so the framework's output will serve to constrain available relation labels rather than describe a single one. For reference, BEFORE constrains the set to TimeML BEFORE or IBEFORE; AFTER to TimeML AFTER or IAFTER and OVERLAPPING to the remaining TimeML relations. An output of VAGUE offers no constraint at all.

TimeML's tense and aspect values were converted to Reichenbachian tenses using the schema given in Table 6.3. These Reichenbachian tenses were then used to find an R-E and an S-R ordering. These orderings for each verb were then coupled, assuming the *R* point for both verbs was shared, in order to determine an ordering between event times. Sometimes this was not possible (e.g. if both are simple past, while both can be described relative to the speech point, they cannot be described with any precision relative to the other); in this case, event orderings were made while falling back to assuming at least a shared *S* point. In other cases, sometimes only a vague relation was possible (e.g. if both are simple present, then they have both happened at some time – speech time – but we know nothing about their starts or ends relative to one another).

Table 6.4 details how constraints were selected. These constraints are translated to TimeML as follows:

Table 6.3 Minimal schema for mapping TimeML event tense and aspects to Reichenbach's framework

Tense	Non-perfect	Perfect
PAST	Simple past	Anterior past
PASTPART	Simple past	Anterior past
PRESENT	Simple present	Anterior present
PRESPART	Simple present	Anterior present
FUTURE	Simple future	Anterior future

Table 6.4 Event orderings based on the Reichenbachian tenses that are available in TimeML. Cell values describe the `e1` [*rel*] `e2` relationship. Note that TimeML has no unambiguous representation for anterior tenses, and so rows for these are not shown

e1 ↓; e2 →	Sim past	Pos past	Ant pres	Sim pres	Ant fut	Sim fut
Sim past	vague	after	vague	after	after	after
Pos past	before	vague	vague	vague	after	after
Ant pres	vague	vague	vague	after	vague	after
Sim pres	before	vague	vague	overlap	vague	after
Ant fut	before	before	vague	vague	vague	after
Sim fut	before	before	before	before	before	vague

Table 6.5 Accuracy of Reichenbach's framework with a subset of links manually annotated for being tensed verbs in the same temporal context

Output	Count	Consistent	% consistent
After	14	4	28.6 %
Overlap	19	15	84.2 %
Before	45	12	26.7 %
Total	78	31	39.7 %
Vague	68	–	–

- `before` - IBEFORE, BEFORE;
- `after` - IAFTER, AFTER;
- `overlap` - everything not covered by `before` or `after`;
- `vague` - no constraint.

As can be seen from prevalence of `vague` entries in the table, many combinations of tense offer no helpful constraint in terms of Allen's interval temporal relations. This is a hint that this particular interpretation of Reichenbach's tense may not see great performance increases when used for relation typing, and (depending on the actual distribution of tenses in the corpus) may not give a very clear picture of how accurate Reichenbach's model is.

The results are in Table 6.5. Indeed, it seems that, using this minimal interpretation, while in some cases Reichenbach's framework generates a temporal ordering that agrees with the TimeBank annotation, in the majority of situations the gold standard temporal orderings are inconsistent with what the framework interpretation suggests (i.e. the suggestion is wrong), or – almost half the time – the framework does not suggest anything useful (e.g. a "vague" response).

6.4.1.1 Minimal Interpretation Failure Analysis

Such low performance from a reasonable framework and interpretation demands analysis. Manual examination of the error set revealed many cases that Reichenbach's framework has problems with.

No Progressive

The framework doesn't handle the progressive aspect. If events have differing tenses (e.g. present and then future), the framework suggests by means of transitivity that the event time of the present-tensed verb is before that of the future-tensed verb. This makes this implicit assumption that the present-tensed item will have completed before the future-tensed item begins, ruling out any possibility of overlap. Progressive aspect is used as an indicator of ongoing processes, and could be used to weaken the constraint imposed by this minimal interpretation. For example, in *"I am running. Heston will cook."*, it is not certain that I will have finished running before the point that Heston starts cooking; that is to say, overlap is possible.

Poor Handling of Long-Running Events

The relations between S, E and R are over-specific information when discussing ongoing events. For example, in *"she hates us and always has hated us"*, a verb is described during another one, but there is a strong tense and aspect shift, from *hates* to *has hated*. Despite looking like a clear example of event ordering, the *hates* is a state that persists, and the speaker is just describing earlier points in the state's existence. However, this interpretation suggests that *hates* is simple present, $S = R = E$, and *has hated* is anterior present, $E < R = S$. This suggests that the event time of *hates* is after that of *has hated* when this is not actually the case. So, in this instance, Reichenbach's framework provides an over-specific response. Although an interpretation of *hates* as a proper interval immediately after the end of *has hated* is not impossible, it is somewhat tenuous, and the facts are too vaguely described to be as certain as the framework is.

Unusual Use of Tense

News presenters do unusual things with tense, and apply the reference point in a flexible manner. In *"And just last month, an off duty policeman is killed when a bomb explodes at another abortion clinic."* The meaning is clear, but the tenses do not compare well with a positional use of the reference point from the *last month* timex. The use of present tense suggests that the passive *killed* and the *explodes* events happen at the same time as the utterance. However, the present tense according to Reichenbach's framework suggests speech and reference time are equal, and in this case, the timex *last month* places speech time explicitly in the month previous to speech time – a direct conflict with the tense framework.

6.4.2 Advanced Interpretation of Reichenbach's Framework

The interpretation of Reichenbach's framework described above makes a few simplifications, and the results are poor. These simplifications may be the cause of incongruence between the framework's apparent suggestions and human-annotated ground-truth data. We improve the interpretation of Reichenbach's framework in the following ways, and re-check it. Some of this section's material also appears in [29].

Account of progressive aspect: In TimeML, aspect values are composed of two "flags", `perfective` and `progressive`, which may both be asserted on any tensed verb. Which Reichenbach's basic framework provides an account of the perfect (which TimeML calls perfective), it does not do the same for the progressive. This is resolved by splitting the event time E into start and finish points E_s and E_f between which the event obtains, as also done by e.g. [30]. For the simple tenses (where $R = E$), described as having TimeML aspect of NONE, it is assumed not that the event is a point, but that the event is an interval (just as in the progressive) and the reference time is *also* an interval, starting and finishing at the same times as the event (e.g. $R_s = E_s$ and $R_f = E_f$).

Variations of context assignment: Reichenbach's definition of which verbs may be linked through permanence of the reference point is a little vague, described as those that share a common reference point. This is approximated in a number of ways, results of each of which are presented: by considering all verb events in the same sentence; by considering all verb events in the same or an adjacent sentence; and by considering all verb events that have a common arrangement of both speech and reference time (e.g. all have the same arrangement of S and R). Ideally one should like to be able to track the speech and reference point through discourse, accounting for relative clauses, embedded phrases, reported speech and the like; in absence of a concerted investigation into performing these tasks reliably automatically, these approaches are approximations.

How to map TimeML to Reichenbach: Instead of the initial approach of mapping the TimeML tense and aspect values to a specific S/R/E point structure (e.g. a relative arrangement of speech, reference and event points) via one of the nine basic tenses specified in Reichenbach's framework, the TimeML tenses and aspects are mapped directly to S/R/E structures, using the translations shown in Table 6.6. For simplicity, PERFECTIVE_PROGRESSIVE aspect was converted to PERFECTIVE; the value makes up for 20 of 5974 verb events, or 0.34 % – a minority that should not have a great impact on overall results if altered slightly. One other simplification is that the participle "tenses" in TimeML (PASTPART and PRESPART) are interpreted in the same way as their non-participle equivalents, and so are not listed.

How to interpret relations suggested by the framework: Previously a label from one of four classes (before, after, overlap, vague) was assigned to a temporal relation, based on the tenses of its participant verb events. These classes did not accurately capture the 14 TimeML relations, and in many cases represented a disjunction of possible interval relation types. Working on the hypothesis that Reichenbach's framework may constrain a TimeML relation type to more than just four possible

Table 6.6 TimeML tense/aspect combinations, in terms of the Reichenbach framework

TimeML tense	TimeML aspect	Reichenbach structure
PAST	NONE	$E = R < S$
PAST	PROGRESSIVE	$E_s < R < S, R < E_f$
PAST	PERFECTIVE	$E_f < R < S$
PRESENT	NONE	$E = R = S$
PRESENT	PROGRESSIVE	$E_s < R = S < E_f$
PRESENT	PERFECTIVE	$E_f < R = S$
FUTURE	NONE	$S < R = E$
FUTURE	PROGRESSIVE	$S < R < E_f, E_s < R$
FUTURE	PERFECTIVE	$S < E_s < E_f < R$

Table 6.7 Example showing disjunctions of TimeML intervals applicable to describe the type of relation between A and B given their tense and aspect (e.g. to describe A rel B)

A ↓ B →	Perfect past	Present progressive
Perfect past	[any]	[before, ibefore, is_included, begins, during]
Present progressive	[after, iafter, includes, begun_by, during_inv]	[simultaneous, identity, during, during_inv, includes, is_included, ends, begins, ended_by, begun_by]

groupings, the table of tense-tense interactions is rebuilt, giving for each event pair a disjunction of TimeML relations instead of one of four labels. This has the advantage of adding distinctions that the minimal framework could not capture. Examples 30 and 31 would both be labeled "before" under that scheme, even though the latter is ambiguous regarding whether the progressive event has finished, and could signify an overlap.

Example 30 Anne had eaten breakfast. Bernard will sing.

Example 31 Chris was cleaning windows. Diana will sleep.

In this case, Example 30 suggests the TimeML relation *eaten* BEFORE *sing*, whereas because the end point of *cleaning* is not certain in Example 31, any of BEFORE, INCLUDES, or ENDED_BY may apply between *cleaning* and *sleep*. In this way, and with other arrangements of the speech, event and reference time, resolving relation types to disjunctions of potential interval relations provides a richer, more descriptive and more precise way of capturing the framework's output. An example is given in Table 6.7.

When constructing a table of potential TimeML TLINK `relType` values given two Reichenbachian tense structures with a disjunction of possible TimeML interval relation types in each cell, there is a finite set of combinations of relation types. That is to say, the disjunctions of interval relations indicated by various tense/aspect pair combinations frequently recur, and are not unique to each tense/aspect pair combination.

This finite set of interval relation disjunctions overlaps with the relation types grouped by Freksa (Sect. 3.2.3). For example, for two events E_1 and E_2, if the tense arrangement suggests that E_1 starts before E_2 (for example, E_1 is simple past and E_2 simple future), the available relation types for E_1/E_2 are BEFORE, IBEFORE, DURING, ENDED_BY and INCLUDES.

To clarify, given that $E_{1s} < E_{2s}$, and $E_s < E_f$ for any proper interval event (e.g. its start is before its finish), the arrangement of E_1 and E_2's finish points is left unspecified. The disjunction of possible interval relation types is as follows:

Table 6.8 Freksa semi-interval relations; adapted from Freksa (1992)

Relation	Illustration
X is *older* than Y Y is *younger* than X	XXX???? YY
X is *head to head* with Y	XXX?? YYYY
X *survives* Y Y is *survived by* X	????XXX YY
X is *tail to tail* with Y	??XXX YYYY
X *precedes* Y Y *succeeds* X	XXX? YYY
X is a *contemporary* of Y	?XXX??? ???YYY?
X is *born before death* of Y Y *dies after birth* of X	XXX????? ?????YYY

- $E_{1f} < E_{2s}$: before;
- $E_{1f} = E_{2s}$: ibefore;
- $E_{1f} > E_{2s}$, $E_{1f} < E_{2f}$: during;
- $E_{1f} = E_{2f}$: ended_by;
- $E_{1f} > E_{2f}$: includes.

In each case, these disjunctions correspond to the Freksa semi-interval relation E_1 YOUNGER E_2. As these Freksa semi-interval relations can be defined in terms of certain groups of Allen relations, the TimeML relations are almost equivalent to the Allen relations and the disjunctions of relations match these TimeML groups perfectly, the "output" of the Reichenbach framework regarding permanence of the reference point is given in Freksa semi-interval relations. The relations are shown in Table 6.8 and the TimeML tense/aspect interaction in Table 6.9.

Results

Interpreted in this way, Reichenbach's framework is more consistent with TimeBank than the earlier, minimal interpretation, generally supporting the framework's suggestions of event-event ordering among pairs of tensed verb events. Results are given in Table 6.10. In this table, an "accurate TLINK" is one where the relation type given in the ground truth is a member of the disjunction of relation types suggested by this interpretation of Reichenbach's framework.

Separate figures are provided for performance including and excluding cases where the disjunction of all link types (e.g. no constraint) is given. This is because achieving consistency with "no constraint" gives no information.

Table 6.9 TimeML tense/aspect pairs with the disjunction of TimeML relations they suggest, according to this chapter's enhanced interpretation of Reichenbach's framework

e1 ↓ e2 →	PAST-NONE	PAST-PROG	PAST-PERF	PRESENT-NONE	PRESENT-PROG	PRESENT-PERF	FUTURE-NONE	FUTURE-PROG	FUTURE-PERF
PAST-NONE	*all*	contemporary	succeeds	survivedby	survivedby	all	precedes	survivedby	before
PAST-PROGRESSIVE	contemporary	*contemporary*	survives	older	all	all	older	born before death	older
PAST-PERFECTIVE	precedes	survivedby	*all*	precedes	survivedby	precedes	before	survivedby	before
PRESENT-NONE	survives	younger	succeeds	*contemporary*	contemporary	survives	precedes	older	older
PRESENT-PROGRESSIVE	survives	all	survives	contemporary	*contemporary*	survives	older	born before death	older
PRESENT-PERFECTIVE	all	all	succeeds	survivedby	survivedby	*all*	before	survivedby	before
FUTURE-NONE	succeeds	younger	after	succeeds	younger	after	*all*	contemporary	survivedby
FUTURE-PROGRESSIVE	survives	dies after birth	survives	younger	dies after birth	survives	contemporary	*contemporary*	survives
FUTURE-PERFECTIVE	after	younger	after	younger	younger	after	survivedby	survivedby	*all*

Table 6.10 Consistency of temporal relation types suggested by Reichenbach's framework with ground-truth data. The non-all column refers to the number of incidences in which there was some kind of relation constraint, e.g., the framework did not give an unhelpful "all relation types possible" response

Context model	TLINKs	Accurate (%)	Non-"all"	Accurate (%)
None (all pairs)	1 167	81.5	481	55.1
Same sentence, same SR	300	88.0	95	62.1
Same sentence	600	71.2	346	50.0
Same/adjacent sentence, same SR	566	91.9	143	67.8
Same/adjacent sentence	913	78.3	422	53.1

Temporal context is complex to automatically detect, as detailed in Sect. 6.3.4 above. These results focus on the accuracy of the framework's temporal relation type constraints, given varying interpretations of temporal context.

The "same SR" context refers to modelling of temporal context as a situation where the ordering of reference and speech times remains constant (in terms of one preceding, occurring with or following the other). The rationale for this temporal context model is, because permanence of the reference point requires a shared reference time, for tenses to be meaningful in their context, the speech time must remain static. This simple same-ordering constraint on S and R does not preclude situations where speech or reference time move, but still remain in roughly the same order (e.g. if reference time moves from 9pm to 9.30pm when speech time is 3pm), which are in fact changes of temporal context (either because R is no longer shared or because S has moved).

In general, consistency is better than with the minimal interpretation discussed above. The "same SR" context gives good results, though has limited applicability in that it considers comparatively reduced sets of TLINKs (e.g. only half of same-sentence links). As both arguments having the same S and R occurs when they have the same TimeML tense, the only variant in these cases – in terms of data that contributes to Reichenbachian interpretation – is the TimeML aspect value. The increased "coverage" of the framework when given the constraint that TLINKs in which both arguments have the same TimeML tense hints that this is a critical factor in interpreting tense, and considering it may lead to improvements in temporal relation typing techniques that rely on aspect, such as that of [31]. The overall result is that Reichenbach's framework is capable of suggesting helpful relation types in some situations, and suggests further effort in applying and using the framework.

A slightly extended, standalone version of this validation can be found in [29].

6.5 Applying Reichenbach's Framework to Temporal Relation Typing

TimeML provides some of the information that Reichenbach's framework alone does not cater for. A combination of the two may lead to better labelling performance, but relying on Reichenbach's framework for rule-based temporal relation label constraint is insufficient. Application of the suggestions as integrated into a machine learning approach is discussed in the next section.

Reichenbach's framework for tense can be used to help determine the relation type between some times and events. This section describes use of the framework to develop features for enhancing temporal relation typing performance. These features are then added to the basic set defined in Sect. 4.4 as part of a temporal relation labelling classifier. The situations we examine are those where two verb events occur in the same temporal context, where a timex directly influences a verb event, and also verb events that report other verb events. A list of features is repeated below.

- text for each event;
- TimeML tense for each event;
- TimeML aspect for each event;
- modality for each event;
- cardinality for each event;
- polarity for each event;
- class for each event;
- part-of-speech for each event;
- are events in the same sentence?;
- are events in adjacent sentences?;
- do events have the same TimeML aspect?;
- do events have the same TimeML tense?;
- does event 1 textually precede event 2?

Because the framework relies on verb tense, all the situations described in this chapter can only work with events that are verbs and with time-referring expressions (that is, TIMEX3s of type DATE or TIME). It is therefore important to correctly determine the subset of all TLINKs that we try relation typing upon. Note that this subset selection is not the same as the relation identification task. The relation identification task requires, given a set of event and timex notifications, the selection of pairs that are temporally related. In contrast, for these experiments it is required, given a set of event, timex and TLINK annotations, to determine which of the TLINKs might benefit from the application of Reichenbach's framework. The relations covered are those that link same-context verbal events, that link events to times, and that link reporting events with events in reported speech. Throughout, the gold-standard EVENT and TIMEX3 annotations found in TimeBank are used, as well as the TLINKs identified there; the only task addressed is that of temporal relation typing.

6.5.1 Same Context Event-Event Links

The framework provides information for determining the ordering of events in the same temporal context (same context event-event links, or the SCEE dataset).

This situation applies to any two verb events that have a shared reference point. Verb events are identifiable by the event having a TimeML POS attribute of VERB, excluding those with a `tense` of `NONE` or `INFINITIVE`. A shared reference point is assumed for all verbs in the same sentence. Sentences are split using the Punkt sentence tokeniser for English [32]. These experiments use the minimal interpretation of Reichenbach's framework, described above.

One new feature is added to the standard feature set, corresponding to the relation type constraint suggested by our advanced interpretation of Reichenbach's framework (Sect. 6.4.2). The only ambiguity is over how to model temporal context. In this case, it is approached as being either event-event links with both arguments in the same sentence, or event-event links with both arguments in the same or adjacent sentences.

6.5.1.1 Results

The experiment was conducted with 10-fold cross validation, considering links from TimeBank v1.2, using relation type folding. The links within a document were never shared across a split (i.e., splits were made at document level). The experiments were conducted with relation folding (see Sect. 3.3.1). The impact of the new feature is measured by comparing classifier performance on SCEE links using the basic feature set and using the basic feature set plus the new feature. Features representing the text (i.e. lexical form) of events were removed as they consistently harmed performance, likely due to the sparsity of their values. Because the splits are determined randomly for cross-fold validation, every experiment is run three times and the mean performance figures given. The results are shown in Table 6.11, and a graph in Fig. 6.2. In this instance, the extended features provide a performance boost regardless of classifier choice.

Table 6.11 Using Reichenbach-suggested event ordering features representing permanence of the reference point, considering only same-sentence TLINKs. 562 examples

Classifier	Base features		Extended features	
	Accuracy (%)	Err. reduction (%)	Accuracy (%)	Err. reduction (%)
Baseline (MCC)	48.04	–	48.04	–
Maxent (megam)	57.47	22.86	**57.65**	23.19
Decision tree (ID3)	56.52	21.14	**57.47**	22.86
Naïve bayes	58.31	24.37	**58.72**	25.12

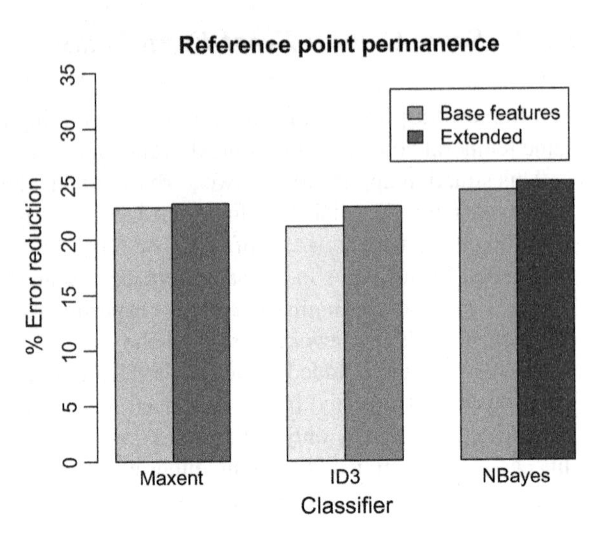

Fig. 6.2 Error reduction in SCEE links with and without features representing permanence of the reference point, modelling temporal context as same-sentence. The darker coloured columns correspond to error reduction using the feature derived from advanced interpretation of Reichenbach's framework

Table 6.12 Reichenbach-suggested event ordering feature representing permanence of the reference point. 858 examples

Classifier	Base features		Extended features	
	Accuracy (%)	Err. reduction (%)	Accuracy (%)	Err. reduction (%)
Baseline (MCC)	44.87	–	44.87	–
Maxent (megam)	62.28	31.58	**62.55**	32.07
Decision tree (ID3)	**59.21**	26.01	58.74	25.16
Naïve bayes	56.96	21.92	**57.58**	23.05

In the next case, the scope of temporal context is broadened to include cases where events are in adjacent sentences. Results are shown in Table 6.12. Here, the classifiers in which inductive bias tends toward the independence assumption do better with the extended feature set, but the decision tree does worse.

In both cases, there was a small performance increase from almost all classifiers with the introduction of the feature derived from advanced interpretation of Reichenbach's framework. Although the gains are not large, they are consistent.

Further work would concentrate on better discriminating which cases can be considered for application of permanence of the reference point. These are likely to span sentences. An annotation for delimiting these cases (e.g. temporal contexts) is put forward later, in Sect. 6.6.

6.5.2 *Same Context Event-Timex Links*

Reichenbach's framework provides explicit rules regarding the rôle of dates and times in respect to a verb within their temporal context (same context event-timex links: SCET). In these cases, the given time determines the time of the reference point, essentially reifying it (see Sect. 6.3.3).

To investigate whether constraints suggested by Reichenbach's framework can help in TLINK relation typing, we proceed as follows. For any verb event that is in the same sentence as a timex, if the timex modifies the event and the timex and event are linked through a TLINK, we assume that the timex positions the verb's reference point, and add a feature corresponding to this.

In all, 684 of the 6 418 available TLINKs could have this principle applied to them (10.7 % of all TLINKs). We are only interested in event-time links, of which there are 2 797; out of this set, 24.5 % (684) have event and time in the same sentence.

6.5.2.1 Features

One new feature is added to the base set (Sect. 4.4). As we are linking a timex and event under the assumption that there is a positional use of the reference point, the reference point is considered equivalent to the timex, and so the interesting temporal ordering is that between R and E. The reference point is determined using the advanced interpretation (Sect. 6.4.2, and the TimeML relation type between R and E constrained using Table 6.4 accordingly. In fact, as can be seen in Table 6.2, the type of tense embodies the E/R ordering: anterior tenses have $E < R$, simple tenses have $E = R$ and posterior tenses have $E > R$. Thus our symbolic label determining E/R relation (which is also E/T relation) assumes the value *anterior*, *simple* or *posterior*.

Dependency parses (generated by the Stanford Parser [33]) help determine whether or not a timex and event are syntactically connected. These parses also yield some extra information, which is included as features. These are:

- Direct modification: Does the timex directly modify the event? E.g., is the timex on the same dependency path as the event? (boolean);
- Temporal modification function: Is there a `tmod` relation in the dependency path from event to timex? (boolean);
- Final relation: The Stanford dependency relation of the timex node and its parent.

6.5.2.2 Results

Experiments were conducted with 10-fold document-level cross validation, using a folded relation set and no lexical features. Each experiment was run three times, and the mean result is reported (Fig. 6.3).

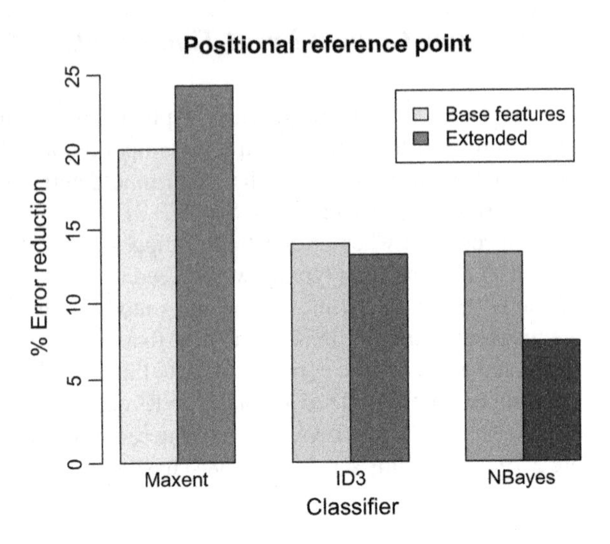

Fig. 6.3 Comparative performance on labelling event-time links where the time positions the reference point

Table 6.13 Performance when using dependency parse and Reichenbach-derived feature, in terms of relation typing accuracy and error reduction above the baseline. 684 instances

Classifier	Base features		Dep. features		RBach features		Dep. + RBach	
	Accuracy (%)	ER (%)	Accuracy (%)	ER (%)	Accuracy (%)	ER (%)	Accuracy (%)	ER (%)
Baseline (MCC)	66.67	–	66.67	–	66.67	–	66.67	–
Maxent (megam)	**73.39**	**20.18**	**74.71**	**24.12**	**74.75**	**24.24**	**74.76**	**24.26**
Decision tree (ID3)	71.35	14.04	70.03	10.09	71.05	13.16	71.10	13.31
Naïve bayes	71.15	13.45	69.74	9.21	70.57	11.69	69.25	7.75

Results are given in Table 6.13. The extended features offered a performance improvement from 20.18 % error reduction to 24.26 % error reduction for the best-performing classifier (maxent). Performance with just the Reichenbach E/R determining feature are also included in the table. The feature is not as useful on its own as it is with the three other dependency-graph derived features.

The absolute increase in labelling accuracy in this subset of TLINKs is approximately 1.4 %; a modest gain, corresponding to an error reduction of. As with investigation into exploiting permanence of the reference point, problems lie in correctly identifying which of the links the features can be applied to.

6.5.3 Summary

Reichenbach's framework for tense and aspect is intuitive, and of moderate utility in typing temporal relations based on the advanced interpretation proposed above. This interpretation has already been shown to be of use when constraining TimeML interval relation types. The big question that remains is about temporal context, which has been only approximated throughout.

The framework suggests helpful constraint in cases where verbs and timexes are in the same context, already helping in automatic relation typing. However, automatic identification of where the framework applies (e.g. temporal contexts) is difficult; this is information not provided in TimeML and not trivially extractable from natural language text. An extended examination of the problems is given in [34].

As the framework is capable of capturing things that TimeML cannot and its utility can be demonstrated in controlled circumstances, it is worth investigating an extension to TimeML to improve on the standard's expressiveness by integrating ideas from Reichenbach.

6.6 Annotating Reichenbach's Framework

Existing temporal annotation schemata are not rich enough to represent all the information in Reichenbach's framework. Critically, although the framework is of use in relation typing, as demonstrated both in this book and also in recent prominent research [35], it cannot be *reliably* applied (and certainly not optimally applied) without knowledge of temporal context. In order to understand temporal context, and move towards using Reichenbach's framework effectively in temporal relation typing, this section details an annotation schema for the framework. Hopefully, given an annotation scheme, it may be possible to annotate text for temporal context and Reichenbachian tense linkages. Having annotations of temporal context enables an investigation into automatically assignment of temporal context, either by plainly revealing the rules that govern where and how contexts start and end, or by providing training data for machine learning approaches.

The new schema proposed for annotating this information is RTMML (Reichenbach Tense Model Markup Language). Following the description of the schema, we introduce a new language resource – a corpus annotated with RTMML. Finally, we demonstrate how it may be integrated with TimeML.

The annotation schema RTMML is intended to describe the verbal event structure detailed in [2], in order to permit the relative temporal positioning of reference, event, and speech times. A simple approach is to define a markup that only describes the information that we are interested in, and can be integrated with TimeML. For expositional clarity we use our own tags but it is possible (with minor modifications) to integrate them with TimeML as an extension to that standard.

Our goal is to define an annotation that can describe S, E and R (speech, event and reference points) throughout a discourse. The lexical entities that these times are attached to are verbal event expressions and temporal expressions. Therefore, our annotation needs to reference these entities in discourse.

6.6.1 Motivation for Annotating the Framework's Points

Critical to knowing how to apply Reichenbach's framework is the issue of temporal context (Sect. 6.3.4). TimeML does not provide an annotation for this phenomenon, and so one must be introduced if we are to develop data to help understand temporal context.

Further, Reichenbach's framework also distinguishes some tenses that are ambiguous in TimeML. Given the 24 permutations for S, E, R and their relations (taken from $<$, $>$, $=$), there are 13 distinct forms, which can be further divided into tenses as below:

- Six arrangements where both relations are $=$ can be boiled down to one, through transitivity of the equality operator. $(24 - 5 = 19)$
- For the twelve arrangements where one relation is $=$, we halve the number of relations that we have, as the ordering of the pair of points connected by $=$ is irrelevant; for example, $S < E = R$ and $S < R = E$ are equivalent. $(19 - 6 = 13)$
- All arrangements where both relations are $<$ are unique and semantically distinct. $(13 - 0 = 13$ tenses)

TimeML's `aspect` attribute will inform us if the reference time is after the event time; that is, if the event is "complete" (to gloss over linguistic nuances detailed by [36]) before the time of reference point. This distinguishes two classes; TimeML `aspect:PERFECTIVE` corresponds to $E < R$, and `aspect:NONE` corresponds to $E \not< R$ (that is, a conflation of $E = R$ and $R < E$).

Also, TimeML does not address the issue of annotating Reichenbach's tense framework with the goal of understanding reference time or creating resources that enable detailed examination of the links between verbal events in discourse. Although other promising solutions are starting to emerge for detailed annotated of tense internals [37], it is not yet possible to describe or build relations to reference points at all in TimeML.

6.6.2 Proposed Solution

Here we discuss what should be annotated in order to capture the information described by Reichenbach's framework, and put forward an annotation schema. Some of this section's material overlaps with [38].

6.6.2.1 Requirements

A schema should allow description of the relations between the three abstract points, speech, reference and event. It must also be capable of expressing relations between different verbs' three points. Finally, it should permit events to be linked with times.

It is preferable to have a schema that follows set frameworks for linguistic annotation, hence supporting interoperability. Hopefully, this can also provide some basic structure for referencing strings within a document and an overall annotation scheme (e.g. XML).

6.6.2.2 Annotation Schema

The annotation language we propose is called RTMML, for Reichenbach Tense Model Markup Language. It includes definitions for document structure and metadata, for verb annotation, for time-referring expression annotation, and for temporal between a verb's three time points.

RTMML documents use standoff annotation. This keeps the text uncluttered, in the spirit of *ISO LAF*[4] and *ISO SemAF-Time*.[5] Annotations reference tokens by their position in the source. Token indices begin from zero. We explicitly state the segmentation plan with the `<seg>` element, as described in [39] and *ISO DIS 24614-1 WordSeg-1*.

The general speech time of a document is defined in the `<doc>` element, which has one optional attribute, `@time` (the `@` indicating that `time` is an attribute name). This is either the string `now` or a normalised value, formatted according to TIMEX3 [40] or TIDES [41].

Each `<verb>` element describes a tensed verb group – that is, a sequence of main and auxiliary verbs that comprise a single verb event. The `@target` attribute describes the verb or group's extents, using segment offsets. It has the form `target="#token0"` or `target="#range(#token7, #token10)"` for a 4-token sequence. Comma-separated lists of offsets are valid, for situations where verb groups are non-contiguous. Every `verb` has a unique value in its `@id` attribute. The Reichenbachian tense structure of a verb group is described using the attributes `@view` (with values *simple*, *anterior* or *posterior*) and `@tense` (*past*, *present* or *future*).

The `<verb>` element has optional attributes for directly linking a verb's speech, event or reference time to a time point specified elsewhere in the annotation. These are `@s`, `@e` and `@r` respectively. To reference the speech, event or reference time of other verbs, we use hash references to the event followed by a dot and then the character `s`, `e` or `r`; e.g., `v1`'s reference time is referred to as `#v1.r`. As well as relating to other verbs, one can reference document creation time with a value of `doc` or a temporal expression with its id (for example, `t1`).

---

[4]ISO 24612:2012 Language resource management – Linguistic annotation framework (LAF).

[5]ISO 24617-1:2012.

Each tensed verb has exactly one S, E and R. As these points do not hold specific values or have a position on an absolute scale, we do not attempt to directly annotate them or assign scalar values to them, instead annotating the type of relation that holds between them. For simplicity, the schema does not split E into incipitive and concluding points (these may still be expressed using TimeML if the two schemas are used in parallel).

One might think that the relations should be expressed in XML links; however this requires reifying time points. The important information is in the relations between Reichenbachian time points, with the actual temporal location of each point often never known. For this reason, the markup focuses on the relations between the Reichenbachian points for each `<verb>`, instead of attempting to assign any kind of value to individual points.

To capture these internal relations for a single verb, we use the attributes `@se`, `@er` and `@sr`. These attributes take a value that is a disjunction of $<$, $=$ and $>$ (though $<$ and $>$ are mutually exclusive). For example, `se=">"` expresses that speech time is after (succeeds) event time.

Time-referring expressions are annotated using the `<timerefx>` element. This has an `@id` attribute with a unique value, and a `@target`, as well as an optional `@value` which works in the same way as the `<doc>` element's `@time` attribute.

6.6.3 Special RTMLINKs

The `<rtmlink>` element is used to connect the speech, reference or event times between given groups of verbs. This is used, for example, for defining a temporal context between verbs that have the same reference time, or annotating positional use of the reference point where a given timex described the reference point of a particular verb event.

To simplify the annotation task, RTMML permits an alternative annotation with the `<rtmlink>` element. The `<rtmlink>` annotation can be used to describe verbs affected by permanence of the reference point (e.g. to reify temporal contexts), positional use of the reference point and positional use of the speech point. This element takes as arguments a relation and a set of times and/or verbs. Possible relation types are POSITIONS, SAME_TIMEFRAME (annotating permanence of the reference point) and REPORTS for reported speech; the meanings of these are given in Table 6.14.

Table 6.14 RTMML relation types

Relation name	Description	Interpretation
POSITIONS	Reference point is set by a timex	$T_a = R_b$
SAME_TIMEFRAME	Verbs in the same temporal context	$R_a = R_b[, R_c, \ldots R_x]$
REPORTS	Reported speech or events	$E_a = S_b$

When more than two entities are listed as `rtmlink` targets, the relation is taken as being between an optional `source` entity and each of the `target` entities. Moving inter-verbal links to the `<rtmlink>` element helps fulfil *TEI p5* and the *LAF* requirements that referencing and content structures are separated.

6.6.4 Example RTMML

This section includes worked examples of sentences and their RTMML annotations.

In Example 32, we define a time *Yesterday* as `t1` and a verbal event *ate* as `v1`.

Example 32
```
<rtmml>
Yesterday, John ate well.
 <seg type="token"/>
 <doc time="now"/>
 <timerefx xml:id="t1"target="
       #token0"/>
 <verb xml:id="v1"target="#token3"
       view="simple"tense="past"
       sr=">"er="="se=">"
       r="t1"s="doc"/>
 </rtmml>
```

The tense of `v1` is placed within Reichenbach's nomenclature, using the `verb` element's `@view` and `@tense` attributes. Next, we directly describe the reference point of `v1`, as being the same as the time `t1`. Finally, we say that this verb is uttered at the same time as the whole discourse – that is, $S_{v1} = S_D$. In RTMML, if the speech time of a verb is not otherwise defined (directly or indirectly) then it is S_D. In cases of multiple voices with distinct speech times, if a speech time is not defined elsewhere, a new one may be instantiated with a string label; we recommend the formatting *s*, *e* or *r* followed by the verb's ID.

This sentence includes a positional use of the reference point, that is, where a time-referring expression determines reference time. This is annotated in `v1` when we say `r="t1"` to verbosely capture a use of the reference point. Further, as the default S/E/R structure of a Reichenbachian simple past tensed verb is non-ambiguous, the attributes signifying relations between time points may be omitted. To simplify the RTMML in Example 32, we could replace the `<verb>` element with that in Example 33:

Example 33
```
<verb xml:id="v1"target="#token3"
      view="simple"tense="past"
      s="doc"/>
<rtmlink xml:id="l1"type="POSITIONS">
  <link source="#t1"/>
```

```
  <link target="#v1"/>
</rtmlink>
```

Longer examples can be found in the appendices, including an excerpt of David Copperfield in Example 34 and Fig. B.1.

6.6.4.1 Comments on Annotation

As can be seen in Table 6.2, there is not a one-to-one mapping from English tenses to the nine specified by Reichenbach. In some annotation cases, it is possible to see from a specific example how to resolve such an ambiguity. In other cases, even if view and tense are not clearly determinable, it is possible to define relations between S, E and R. For example, for arrangements corresponding to the simple future, $S < E$. In cases where ambiguities cannot be resolved, one may annotate a disjunction of possible relation types; continuing the simple future example, we could say "$S < R$ or $S = R$" with `sr="<="`.

Some parts of the annotation task present difficulties. During a trial annotation, while annotators could determine the scoping exercise that is temporal context annotation without too much difficulty, directly mapping a verb group to a single Reichenbachian tense schema was hard, and at best tiring. Decomposing this task into pairwise judgements between S, E and R made annotation easier, though when one could often not see all the information required in order to make the correct judgement; as a result, many pairwise annotations were changed after annotators considered distinct but related pairs. Posing the annotation task as one of temporal constraint, using more concrete ideas (e.g. "From the text, does this event of *John running* obtain at 9p.m.?" instead of "Is T_9 during E_7?") may reduce annotator fatigue and error. RTMML does not address intentionality, leaving this to annotators and, where expressable, TimeML (which includes the I_ACTION and I_STATE event classes for this purpose).

RTMML annotation is also independent of language. As long as a segmentation scheme (e.g. WordSeg-1) is agreed, the model can be applied and an annotation created.

6.6.4.2 Integration with TimeML

To use RTMML as an ISO-TimeML extension, we recommend that instead of annotating and referring to `<timerefx>`s, one refers to `<TIMEX3>` elements using their `tid` attribute; references to `<doc>` will instead refer to a `<TIMEX3>` that describes document creation time. The attributes of `<verb>` elements (except `xml:id` and `target`) may be be added to `<EVENT>` elements, and `<rtmlink>`s will refer to event or event instance IDs.

6.7 Chapter Summary

Previous findings suggested that tense shifts played a significant part in temporal relation typing, especially of difficult links. To this end, in this chapter, we introduced Reichenbach's framework for tense and aspect. The chapter introduced novel additions to the framework, and proposed two interpretations of it (one minimal, one advanced) in the context of TimeML. The advanced interpretation was used to perform the first validation of Reichenbach's framework against gold-standard temporally annotated resources, and provided empirical support for Reichenbach's 65-year-old theoretical framework. While showing support for the framework, the validation also uncovered important issues regarding how to choose which events or times could be linked, which is described in this book as "temporal context".

Given the framework, a method of interpreting it and a demonstration of its validity, this chapter also investigated how to leverage the framework in the overall problem of the relation typing task. Various approaches to using Reichenbach's framework in machine learning approaches to temporal relation typing were described. This allowed experimentation with different approximations of temporal context, and showed that the framework can be leveraged for real temporal relation typing gains.

These empirical results supported a further investigation into temporal context, which is begun with the introduction in this chapter of an annotation schema for Reichenbach's framework, that permits not only delineation of temporal context bounds but also annotation of reference time, as well as speech and event times in a corpus.

References

1. Lapata, M., Lascarides, A.: Learning sentence-internal temporal relations. J. Artif. Intell. Res. **27**(1), 85–117 (2006)
2. Reichenbach, H.: The tenses of verbs. In: Elements of Symbolic Logic. Dover Publications (1947)
3. Harris, R., Brewer, W.: Deixis in memory for verb tense. J. Verbal Learn. Verbal Behav. **12**(5), 590–597 (1973)
4. Hepple, M., Setzer, A., Gaizauskas, R.: USFD: preliminary exploration of features and classifiers for the TempEval-2007 tasks. In: Proceedings of the 4th International Workshop on Semantic Evaluations, SemEval 2007, pp. 438–441. Association for Computational Linguistics (2007)
5. Pratchett, T.: The Light Fantastic. Colin Smythe, Gerards Cross (1986)
6. Matlock, T., Ramscar, M., Boroditsky, L.: On the experiential link between spatial and temporal language. Cogn. Sci. **29**(4), 655–664 (2005)
7. Huggett, N.: Zeno's paradoxes. In: Zalta, E.N. (ed.) The Stanford Encyclopedia of Philosophy (Winter), 2010 edn. CSLI (2010)
8. Eddington, A.: The Nature of the Physical World. Macmillen, Cambridge (1928)
9. Stocker, K.: The time machine in our mind. Cogn. Sci. **36**, 385–420 (2012)
10. McTaggart, J.: The unreality of time. Mind **17**(4), 457 (1908)
11. Lyons, J.: Semantics, vol. 2. Cambridge University Press, Cambridge (1977)

12. Michaelis, L.: Time and tense. The Handbook of English Linguistics, pp. 220–243. Blackwell, Oxford (2006)
13. Jaszczolt, K.M.: Representing Time: An Essay on Temporality as Modality. Oxford University Press, Oxford (2009)
14. Fillmore, C.: Lectures on Deixis. CSLI Publications Stanford, California (1971)
15. Klein, W.: Time in Language. Germanic Linguistics. Routledge, London (1994)
16. Mani, I., Pustejovsky, J., Gaizauskas, R.: The Language of Time: A Reader. Oxford University Press, Oxford (2005)
17. Song, F., Cohen, R.: The interpretation of temporal relations in narrative. In: Proceedings of the 7th National Conference of AAAI (1988)
18. Ahn, D., Adafre, S., Rijke, M.: Towards task-based temporal extraction and recognition. In: Dagstuhl Seminar Proceedings, vol. 5151 (2005)
19. Hornstein, N.: As Time goes by: Tense and Universal Grammar. MIT Press, Cambridge (1990)
20. Dowty, D.: The effects of aspectual class on the temporal structure of discourse: semantics or pragmatics? Linguist. Philos. **9**(1), 37–61 (1986)
21. Miller, T.A., Bethard, S., Dligach, D., Pradhan, S., Lin, C., Savova, G.K.: Discovering narrative containers in clinical text. ACL **2013**, 18 (2013)
22. By, T.: Tears in the rain. Ph.D. thesis, University of Sheffield (2002)
23. Paslawska, A., van Stechow, A.: Perfect readings in russian. Perfect Explorations **2**, 307 (2003)
24. Hinrichs, E.: Temporal anaphora in discourses of english. Linguist. Philos. **9**(1), 63–82 (1986)
25. Giorgi, A., Pianesi, F.: Tense and Aspect: From Semantics to Morphosyntax. Oxford University Press, USA (1997)
26. Hristova, D.: The neoreichenbachian model of tense syntax and the rusian active participles. Harv. Ukr. Stud. **28**(1/4), 155–164 (2006)
27. Prior, A.: Past, Present and Future. Clarendon, Oxford (1967)
28. Tanaka, K.: On reichenbach's approach to tense. Tsukuba Engl. Stud. **9**, 61–75 (1990)
29. Derczynski, L., Gaizauskas, R.: Empirical validation of reichenbach's tense framework. In: Proceedings of the 10th Conference on Computational Semantics, pp. 71–82. Association for Computational Linguistics (2013)
30. Kowalski, R., Sergot, M.: A logic-based calculus of events. In: Foundations of Knowledge Base Management, pp. 23–55. Springer (1989)
31. Costa, F., Branco, A.: Aspectual type and temporal relation classification. In: Proceedings of the 13th Conference of the European Chapter of the Association for Computational Linguistics, pp. 266–275 (2012)
32. Kiss, T., Strunk, J.: Unsupervised multilingual sentence boundary detection. Comput. Linguist. **32**(4), 485–525 (2006)
33. De Marneffe, M., MacCartney, B., Manning, C.: Generating typed dependency parses from phrase structure parses. In: Proceedings of the International Conference on Language Resources and Evaluation (2006)
34. Derczynski, L., Gaizauskas, R.: Temporal relation classification using a model of tense and aspect. In: Proceedings of the Conference on Recent Advances in Natural Language Processing. Association for Computational Linguistics (2015)
35. Chambers, N., Cassidy, T., McDowell, B., Bethard, S.: Dense event ordering with a multi-pass architecture. Trans. Assoc. Comput. Linguist. **2**, 273–284 (2014)
36. Vendler, Z.: Verbs and times. Philos. Rev. **66**(2), 143–160 (1957)
37. Gast, V., Bierkandt, L., Rzymski, C.: Creating and retrieving tense and aspect annotations with GraphAnno, a lightweight tool for multi-level annotation. In: Proceedings 11th Joint ACL-ISO Workshop on Interoperable Semantic Annotation (ISA-11), pp. 23–28 (2015)
38. Derczynski, L., Gaizauskas, R.: An Annotation Scheme for Reichenbach's Verbal Tense Structure. In: Workshop on Interoperable Semantic Annotation, pp. 10–17 (2011)
39. Lee, K., Romary, L.: Towards Interoperability of ISO Standards for Language Resource Management. In: International Conference on Global Interoperability for Language Resources (2010)

40. Boguraev, B., Ando, R.: TimeML-compliant text analysis for temporal reasoning. In: Proceedings of International Joint Conference on Artificial Intelligence (IJCAI) (2005)
41. Ferro, L., Gerber, L., Mani, I., Sundheim, B., Wilson, G.: Tides 2005 standard for the annotation of temporal expressions. Technical report 03–1046, The MITRE Corporation (2005)

Chapter 7
Conclusion

> Yesterday, you said tomorrow. So just do it.
>
> SHIA LEBOEUF

Temporal annotation is difficult for both humans and machines. The task of determining how particular events are ordered or nested is part of this temporal annotation problem and has been the goal of this book. This is known as the temporal link labelling problem. The state of the art in this problem has advanced slowly in recent years, without reaching high enough performance levels to consider it solved. This book has investigated the problem of temporal link labelling.

A principled investigation began with a data-driven exploration of temporal links in a publicly-available corpus. This led to the identification of a set of difficult links, which many modern approaches cannot automatically label correctly. Formal and subjective analyses of this difficult link set were conducted. Results suggested multiple avenues of research (in the form of types of information seemingly used to label temporal links) and the two that were selected for investigation were signal-based links and links where there is a change of tense or aspect.

For the part of the signals, these were characterised as words or phrases associated with a pair of events or timexes that provide explicit information about their temporal relation. Experimentation with a machine learning approach showed that they were very helpful in link labelling, giving about a 50 % error reduction. However, they are under-annotated in TimeBank, so attention turned to the task of automatically annotating signals. This was broken down into a two part task: discriminating signals (e.g. finding which phrases occur in text with a temporal sense and in a link labelling-supportive function) and association of signals, that is, determining which pair of events or timexes has its relation described by a given signal. Machine learning approaches and feature sets were identified for both these tasks. Finally, automatic signal annotation was attempted on a corpus initially devoid of signals and the automatically-found signals used to help classifier-based temporal link labelling on

L.R.A. Derczynski, *Automatically Ordering Events and Times in Text*,
Studies in Computational Intelligence 677, DOI 10.1007/978-3-319-47241-6_7

that corpus, yielding an overall benefit compared to automatic labelling without any signal information.

To address the cases of tense shifts, Reichenbach's framework of tense was investigated. This included multiple interpretations from the framework to TimeML, including various mappings from tense and aspect pairs into its own tense structure. The framework proposes event and time orderings in simple and complex situations, based on a point-wise temporal logic. The framework also includes capacity for expression of abstract temporal points that is not present in TimeML. Initial validation suggested that the model could be of use for constraining the types of temporal relation between a given linked pair of event verbs. The model's output was added as a feature in a machine learning approach for temporal link labelling, and found to be of some utility in most cases. However, the problem of determining which events and times may be linked through this framework is open, and difficult to solve with existing tools. Critically, no existing resources are available in which this "temporal context" is annotated. A markup acting as an extension to TimeML is proposed for supporting this functionality, as well as supporting reasoning with and annotation for other aspects of Reichenbach's framework.

Overall, an investigation began with analysis of difficult temporal relations. Potential sources of information were identified that could be used to improve automatic system's performance when determining the types of these difficult relations. Of these, two were investigated – explicit temporal signals, and tense – and both exploited in such a way as to improve temporal relation typing. In the course of this exploitation a better understanding of discourse temporal relations and of both phenomena was reached, explained within this book.

7.1 Contributions

The work presented in this book furthered the understanding of some mechanisms used to convey temporal information in language.

7.1.1 Survey of Relations and Relation Typing Systems

Chapter 4 contained a data-driven analysis of temporal relation systems, in an attempt to first identify which relations are the hardest to automatically assign types to, and then to analysis this set of "difficult" links. TempEval-2 was an evaluation exercise where many systems attempted temporal relation labelling over a common data set. The exercise comprises the first analysis of the TempEval-2 participants' performance at relation-level, and the most in-depth analysis of any TempEval exercise.

As well as developing a definition of difficult links and defining a set of those links that are the hardest to automatically label within the TempEval-2 corpus, the chapter presents quantitative and qualitative analyses of the difficult link set. In this set, there were large groups of temporal links using explicit signals and others using tense shifts. These phenomena form the basis of the remainder of the book' investigation.

7.1.2 Temporal Signals

Chapter 5 investigated the role of explicit temporal signals in discourse, with regard to temporal relations. This chapter introduced a method for using signals to achieve a large relation typing performance boost on the temporal links that they co-ordinate. Seeing that signals can be useful, a characterisation of signals is presented, as well as a corpus survey of them. Finding under-annotation in TimeBank, temporal signal annotation guidelines are clarified and an augmented version of TimeBank including extra signals (and, as a result, some extra events, timexes and temporal links) is created. Given evidence for the utility of signals and high-quality ground truth data, the chapter turns to the automatic annotation of temporal signals. This annotation task is split into two sub-parts: signal discrimination (distinguishing temporal from non-temporal uses of signal words) and signal association (finding which timexes or events a given signal co-ordinates). Successful automatic methods for independent signal discrimination and signal association are introduced. These two sub-parts are then joined, in a joint annotation approach, and this approach for signal annotation evaluated, with satisfactory results. Finally, the question of the approach's ability to contribute to the overall temporal relation typing task is addressed. The joint approach is used to label signals and connect them to temporal relations. The results indicate an improvement in temporal relation labelling after this chapter's signal annotations are applied to a document.

7.1.3 Framework of Tense and Aspect

Building on the earlier analysis of difficult links, Chap. 6 introduces a theoretical framework for dealing with tense and aspect – that of [1]. This chapter first introduced tense and the framework, and suggested extensions to the framework to account for positional use of the speech point. Before applying the framework to the temporal relation typing task, it was rational to validate it. This was attempted using a minimal interpretation of the framework, with negative results. Failure analysis led to a new, advanced interpretation, including several novel concepts: an account of progressives; the notion of temporal context (groups within which certain tense rules can be applied); and the discovery that event-event relation typing based on tense suggests relations in semi-interval-link groupings. This advanced interpretation led to the first empirical validation of Reichenbach's framework of tense and aspect. Continuing, techniques for integrating the framework in supervised approaches for event-event and event-timex relation typing were introduced, giving slight benefits over the same approaches without information suggested by the framework. Problems were found with accurately automatically determining temporal context; a lack of context detection limits the applicability of the framework. The chapter closed with the description of a markup language for Reichenbach's framework, integrated with a current temporal annotation schema, in order to further research in this demonstrably promising area.

7.2 Future Work

The book suggests many directions of future work throughout. This section highlights some key points.

7.2.1 Sources of Difficult Links

The failure analysis of temporal relation typing given in Chap. 4 suggests a large number of directions for further investigation. Only two of the problem areas discovered are explored in the rest of the book: signals and tense shifts. Many questions are raised about, for example, the impact that modality, iconicity, world knowledge and textual proximity have upon temporal relations. All these linguistic phenomena are worthy of further investigation, so that their rôle in temporal relation typing might be determined.

Recurrent is the theme of inference: the idea that the configuration of some temporal relations has a constraining impact on the possible configurations of other temporal relations. Temporal closure forms the basic part of this concept, but the role of temporal inference still remains largely unexplored. Approaches that attempt to use it often see only small improvements, though because global temporal constraint is difficult to perform, they have only included reduced-scope models of temporal inference. In an area full of noisy supervised learning output, it would be interesting to see a better integration of global temporal constraints. Prior work on temporal constraint networks [2] has come close to this area. Techniques that can integrate the noisy, uncertain classifier output with global temporal constraints and discourse structure may yield new levels of temporal relation typing performance.

7.2.2 Temporal Signals

While Chap. 5 introduced successful approaches for both annotating temporal signals and exploiting them for temporal relation typing, each of these approaches is a prototype and the first of its kind. There is certainly scope for improvement on each front.

Signal discrimination can be seen as a simplified word sense disambiguation (WSD) task: we are distinguishing temporal from non-temporal uses of expressions. While part-of-speech was shown not to be enough to determine whether or not a given signal was temporal, the approach taken still ignores the majority of the WSD literature [3]. For example, no context is taken into account when performing discrimination. Testing state-of-the-art WSD approaches on this binary classification task may lead to interesting results. Perhaps also the signal discrimination approach given in the chapter may contribute to some WSD tasks.

Signal association is a non-trivial task, and the approach given has some intrinsic limitations. For example, with the best-performing approach, only interval pairs within a certain number of sentences of each other are considered. This is shown by data from the corpus to already exclude some relations where the pair of intervals lie far apart. Other approaches to signal association, perhaps incorporating different discourse relations or some knowledge of pragmatics, may remove these boundaries and lead to increased performance.

Spatial and temporal signals are shown to have a lot in common. Spatial signals also seem to be critical in description of some spatial relations. It follows that the approach detailed in this book may be mapped without too much difficulty to the problem of automatically annotating spatial signals, and perhaps even to using them in automatic spatial relation typing [4].

Finally, given the success of the signal annotation approach and the lack of signal annotation capability in current temporal annotation tools (e.g. [5]), a next logical step is to package the techniques developed during the course of this book into a distributable tool for temporal signal annotation.

7.2.3 Reference Time and Temporal Context

The work presented on Reichenbach's framework, and the new evaluation of its validity, progress many existing problems in computational linguistics concerning the management and interpretation of time in discourse. The chapter presents a big problem: that of determining temporal context. Clearly this is a direction for further work, marshalling current progress in discourse segmentation, syntactic analysis and the behaviour of temporal expressions. The results in this chapter suggest that automatically understanding temporal context permits accurate event-event and event-time relation typing.

However, temporal context is not the sole avenue for further research based on Reichenbach's framework. Multiple problems have called for a means of determining and reasoning with reference time. Aside from the temporal relation typing task, timex normalisation (interpreting an expression of a time) and story generation both require nuanced temporal reasoning, including awareness of the reference point.

Some existing temporal expression normalisation systems heuristically approximate reference time. GUTime [6] interprets the reference point as "the time currently being talked about", defaulting to document creation date. Over 10 % of errors in this system were directly attributed to having an incorrect reference time, and correctly tracking reference time is the only way to resolve them. TEA [7] approximates reference time with the most recent timex temporally (as opposed to textually) before the expression being evaluated, excluding noun-modifying temporal expressions; this heuristic yields improved performance in TEA when enabled, showing that modelling reference time helps normalisation. HeidelTime [8] uses a similar approach to TEA but does not exclude noun-modifying expressions.

The model is of use when generating language, for determining which tense to use. In fact, it is necessary to consider abstract temporal entities such as the reference point in order to know when to shift tense and how to properly describe events in other temporal frames of reference. A formal application of the model as it extends TimeML may prove useful to accurate language generation. Elson [9] describes how to relate events based on a "perspective" which is calculated from the reference and event times of an event pair. The authors construct a natural language generation system that requires accurate reference times in order to correctly write stories.

Portet [10] found reference point management critical to medical summary generation, in a situation where many small reports were generated with shifting speech and reference points, in order to helpfully unravel the meanings of tense shifts in minute-by-minute patient reports.

The WikiWars corpus of TIMEX2 annotated text prompted the comment that there is a "need to develop sophisticated methods for temporal focus tracking if we are to extend current time-stamping technologies" [11]. Resources that explicitly annotate reference time will be direct contributions to the completion of this task.

A computational model of the sequence of tenses may offer improvements in automatic machine translations. This is because accurately capturing temporal context permits more precise "analytical interlingual translation" [12].

There is also demand in journalism for changing a stock wire articles between present, past and anterior past, in order to suit a particular outlet's style guidelines. This mood switching can be accomplished using Reichenbach's framework.

Finally, the problem of datestamping documents automatically is not trivial. Reichenbach's framework provides the notion of speech time and means of bounding using permanence of the reference point between same-context events and attachment of events to fixed times via positional use of the reference point with a document's timexes. The model may therefore provide insights into this problem.

In summary, automatic determination of reference time for verbal expressions is an open problem, the solution of which is useful for a number of computational language processing tasks.

References

1. Reichenbach, H.: The tenses of verbs. In: Elements of Symbolic Logic. Dover Publications (1947)
2. Dechter, R., Meiri, I., Pearl, J.: Temporal constraint networks. Artif. Intell. **49**(1), 61–95 (1991)
3. Navigli, R.: Word sense disambiguation: a survey. ACM Comput. Surv. **41**(2), 1–69 (2009)
4. Kordjamshidi, P., Van Otterlo, M., Moens, M.: Spatial role labeling: towards extraction of spatial relations from natural language. ACM Trans. Speech Lang. Process. (TSLP) **8**(3), 4 (2011)
5. Verhagen, M., Mani, I., Sauri, R., Knippen, R., Jang, S., Littman, J., Rumshisky, A., Phillips, J., Pustejovsky, J.: Automating temporal annotation with TARSQI. In: Proceedings of the ACL 2005 on Interactive Poster and Demonstration Sessions, p. 84. Association for Computational Linguistics (2005)
6. Mani, I., Wilson, G.: Robust temporal processing of news. In: Proceedings of the 38th Annual Meeting on Association for Computational Linguistics, pp. 69–76. Association for Computational Linguistics (2000)
7. Han, B., Gates, D., Levin, L.: From language to time: a temporal expression anchorer. In: Proceedings of the 13th International Symposium on Temporal Representation and Reasoning (TIME) (2006)
8. Strötgen, J., Gertz, M.: HeidelTime: High quality rule-based extraction and normalization of temporal expressions. In: Proceedings of the 5th International Workshop on Semantic Evaluation, pp. 321–324. Association for Computational Linguistics (2010)
9. Elson, D., McKeown, K.: Tense and aspect assignment in narrative discourse. In: Proceedings of the Sixth International Conference in Natural Language Generation (2010)
10. Portet, F., Reiter, E., Gatt, A., Hunter, J., Sripada, S., Freer, Y., Sykes, C.: Automatic generation of textual summaries from neonatal intensive care data. Artif. Intell. **173**(7–8), 789–816 (2009)
11. Mazur, P., Dale, R.: WikiWars: a new corpus for research on temporal expressions. In: Proceedings of the EMNLP, pp. 913–922 (2010)
12. Horie, A., Tanaka-Ishii, K., Ishizuka, M.: Verb temporality analysis using Reichenbach's tense system: Towards interlingual MT. In: Proceedings of the International Conference on Computational Linguistics, pp. 471–482. Association for Computational Linguistics (2012)

References

Appendix A
Annotated Corpora and Annotation Tools

A.1 Introduction

TimeML is a standard for annotating time in natural language. It includes annotations for the lexicalised entities TIMEX3, EVENT and SIGNAL, and for the abstract entities TLINK, SLINK, ALINK and MAKEINSTANCE. The syntax is XML-like, with inline annotation. For the temporal link labelling task, one is interested in TIMEX3, EVENT, SIGNAL and TLINK. The MAKEINSTANCE tag gives events extra information and instantiates them for use in TLINKs, and so also contains useful information. TimeML has recently become an ISO standard, ISO-TimeML, which incorporates a few changes to event description and permits stand-off annotation. As almost all prior work and all existing resources use TimeML or an extension thereof, this book considers only TimeML in general.

A.2 Corpora

A.2.1 TimeBank

TimeBank is a human annotated TimeML corpus of 183 newswire texts. TimeBank v1.2 contains 6 418 TLINKs, 1 414 TIMEX3s and 7 935 EVENTs, and is 3004 kB in size. This is tiny compared to some other types of corpus, but is large enough to be useful and has been battered enough by the community through a few versions to be considered robust. TimeBank's creation [1] involved a large human annotator effort and a few different versions [2]; it is currently the largest temporally annotated corpus.

TimeBank 1.2 contains 183 documents, comprising about 64 000 tokens. Over these tokens are:

- 7935 EVENTs
- 6418 TLINKs

L.R.A. Derczynski, *Automatically Ordering Events and Times in Text*,
Studies in Computational Intelligence 677, DOI 10.1007/978-3-319-47241-6

Table A.1 Inter-annotator agreement in TimeBank v1.2; data from [2]

TimeML tag	Exact match IAA
TIMEX3	0.83
EVENT	0.78
TLINK	0.55

Table A.2 Distribution of TIMEX3 type

TIMEX3 type	Frequency	Proportion (%)
DATE	1164	82.3
DURATION	175	12.4
TIME	63	4.46
SET	12	0.849
Total	1414	

Table A.3 Distribution of TIMEX3 mod

TIMEX3 mod	Frequency	Proportion (%)
START	28	30.4
APPROX	16	17.4
END	16	17.4
EQUAL_OR_LESS	8	8.7
MID	7	7.61
EQUAL_OR_MORE	6	6.52
LESS_THAN	4	4.35
MORE_THAN	3	3.26
ON_OR_AFTER	3	3.26
BEFORE	1	1.09
None	0	0.0
Total	92	

- 7940 INSTANCEs
- 688 SIGNALs
- 1414 TIMEX3s
- 2932 SLINKs
- 265 ALINKs

The remainder of this subsection presents summary information over the events, timexes, signals and and temporal relations in TimeBank 1.2 (Tables A.1, A.2 and A.3).

A.2.2 AQUAINT

The second-largest English TimeML corpus is the AQUAINT TimeML corpus. The AQUAINT TimeML corpus consists of around 80 TimeML-annotated newswire documents. These are grouped by the story that they cover, with each group related to the same story, reporting progress on events through time (Table A.4).

Due to repeated text and heavy event co-reference, the AQUAINT corpus requires some care to use correctly. One must firstly maintain document level testing and training set separation, to ensure that evaluation examples are not those found verbatim in training data. Further, due to the corpus' repeated attention to the same story over multiple documents, some event summaries and orderings are repeated using the same text across documents. For this reason, it is best to split datasets by story, so that the background summaries repeated in articles on the same story do not contaminate test and training data. Finally, separately from text re-use, there is re-description of events using later knowledge. Because the news stories contain information on the same topic describing the same events, it is important not to include later articles in the training set for a classifier evaluated on articles published prior. That is to say, evaluation should not be performed using articles that the training data provides hindsight over. This is a common constraint with time-series data [3] and applies to this TimeML corpus because of its repeated coverage of the same super-events (Table A.5).

Table A.4 Distribution of EVENT class

EVENT class	Frequency	Proportion (%)
OCCURRENCE	4215	53.1
STATE	1117	14.1
REPORTING	1028	13.0
I_ACTION	681	8.58
I_STATE	584	7.36
ASPECTUAL	262	3.3
PERCEPTION	48	0.605
Total	7935	

Table A.5 Distribution of EVENT pos

EVENT pos	Frequency	Proportion (%)
VERB	5122	64.5
NOUN	2225	28.0
OTHER	299	3.77
ADJECTIVE	266	3.35
PREPOSITION	28	0.353
Total	7940	

A.2.3 Other TimeML Corpora

There have been other TimeML corpora released, in a range of languages, including French [4], Italian [5] and Romanian [6] (Tables A.6 and A.7).

Table A.6 Distribution of EVENT modality

EVENT modality	Frequency	Proportion (%)
would	127	39.7
could	49	15.3
may	31	9.69
can	26	8.13
none	21	6.56
might	16	5.0
must	14	4.38
should	13	4.06
have to	5	1.56
’d	2	0.625
possible	2	0.625
should have to	2	0.625
close	1	0.313
delete	1	0.313
depending on	1	0.313
have_to	1	0.313
having to	1	0.313
likelihood	1	0.313
potential	1	0.313
to	1	0.313
unlikely	1	0.313
until	1	0.313
would have to	1	0.313
would_be	1	0.313
None	0	0.0
Total	320	

Table A.7 Distribution of EVENT polarity

EVENT polarity	Frequency	Proportion (%)
POS	7651	96.4
NEG	289	3.64
Total	7940	

A.2.4 Other Non-TimeML Corpora

The TempEval corpora [7, 8] feature event, timex and tlink annotations over non-parallel news text in multiple different languages. The set of TLINKs is slightly different from those available in TimeML, being simpler and including a VAGUE relation. TempEval-2 included English, Spanish, French, Italian, Chinese and Korean (Table A.8).

The ACE (Automatic Content Extraction) exercises were based on purpose-built corpora that included a large number of TIMEX2 annotations, comprising almost 26 000 TIMEX2s. For comparison, TimeBank has only 1 414 TIMEX3 annotations (Table A.9).

The WikiWars corpora [9, 10] are derived from WikiPedia articles about wars. These articles tend to contain temporal expressions of a variety of granularities and forms and a generally quite long pieces of connected prose. WikiWars and WikiWars-DE are both annotated according to TIMEX2 and are resources of significant size.

Table A.8 Distribution of TLINK reltype

TLINK reltype	Frequency	Proportion (%)
BEFORE	1408	21.9
IS_INCLUDED	1357	21.1
AFTER	897	14.0
IDENTITY	743	11.6
SIMULTANEOUS	671	10.5
INCLUDES	582	9.07
DURING	302	4.71
ENDED_BY	177	2.76
ENDS	76	1.18
BEGUN_BY	70	1.09
BEGINS	61	0.95
IAFTER	39	0.608
IBEFORE	34	0.53
DURING_INV	1	0.0156
Total	6418	

Table A.9 Transitivity table for the TimeML relation set; an X indicates that no clear inference can be made. Abbreviations: BE = Before, AF = After, IN = Includes, II = Is_included, DU = During, SI = Simultaneous, IA = Iafter, IB = Ibefore, ID = Identity, BG = Begins, EN = Ends, BB = Begun_by, EB = Ended_by, DI = During_inv

A r1 B	B r2 C													
	BE	AF	IN	II	DU	SI	IA	IB	ID	BG	EN	BB	EB	DI
BEFORE	BE	X	BE	X	BE	BE	X	BE	BE	BE	X	BE	BE	BE
AFTER	X	AF	AF	X	AF	AF	AF	X	AF	X	AF	AF	AF	AF
INCLUDES	X	X	IN	X	IN	IN	X	X	IN	X	X	IN	IN	IN
IS_INCLUDED	BE	BE	X	II	II	II	AF	BE	II	II	II	X	X	II
DURING	BE	AF	IN	II	SI	SI	IA	IB	SI	BG	EN	BB	EB	SI
SIMULTANEOUS	BE	AF	IN	II	SI	SI	IA	IB	SI	BG	EN	BB	EB	SI
IAFTER	X	AF	AF	X	IA	IA	AF	X	IA	X	IA	AF	IA	IA
IBEFORE	BE	X	BE	X	IB	IB	X	BE	IB	IB	X	IB	BE	IB
IDENTITY	BE	AF	IN	II	SI	SI	IA	IB	ID	BG	EN	BB	EB	SI
BEGINS	BE	AF	X	II	BG	BG	IA	BE	BG	BG	II	X	X	BG
ENDS	BE	AF	X	II	EN	EN	AF	IB	EN	II	EN	X	X	EN
BEGUN_BY	BE	AF	IN	X	BB	BB	IA	X	BB	X	X	BB	IN	BB
ENDED_BY	BE	X	IN	X	EB	EB	X	IB	EB	X	X	X	EB	EB
DURING_INV	BE	AF	IN	II	SI	SI	IA	IB	SI	BG	EN	BB	EB	SI

A.3 Temporal Annotation Tools

Temporal annotation is a complex task for humans; to this end, we have annotation guidelines to simplify things. Typing XML is also a rather painful experience for us, let alone a specific variant of it that captures abstract information, such as TimeML; and to this end, we have temporal annotation tools that can simplify the task.

In this section, we will first describe TARSQI, a state-of-the-art toolkit containing many components for temporal annotation of text. We will then discuss the problem of visually presenting temporal information.

A.3.1 TARSQI/TTK

A set of tools for automatic TimeML annotation are bundled together in the form of the TARSQI toolkit, TTK [11], which is described as "a modular system for automatic temporal and event annotation of natural language texts" (Fig. A.1). TTK adopts a multi-stage work-flow, beginning with the entry of raw unannotated text, followed by automated annotation and then user correction of machine-produced results. The toolkit ties together a large number of components, including EVITA [12], Slinket [13, 14], SputLink [15] and TBOX [16], using a plugin-based Python framework. It is easy for users to see which plugins have been involved in annotation decisions, making TTK useful for analyzing individual components.

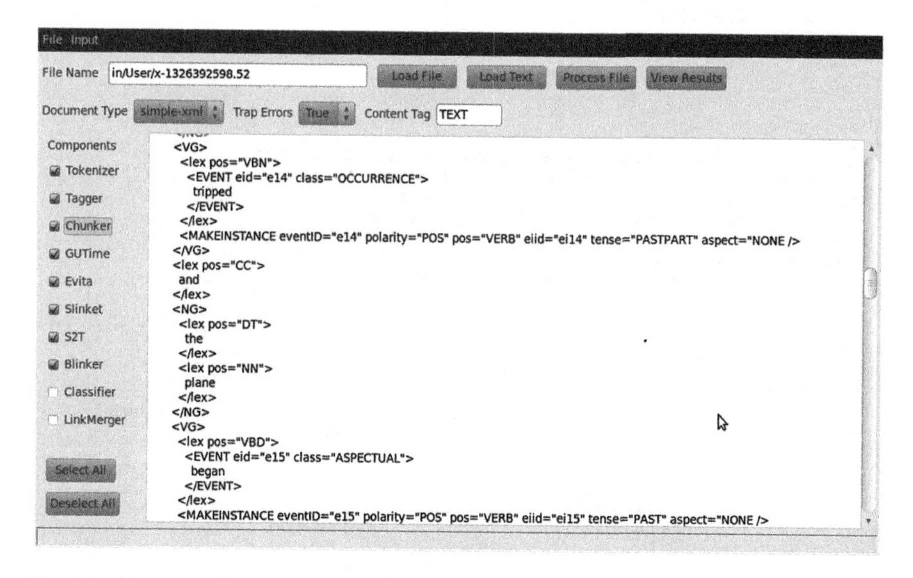

Fig. A.1 Automatically annotating text with TTK

As well as identifying and annotating events and times, TTK also includes sophisticated logic for labelling TLINKs. As far as rule-based relation identification goes, S2T [11, 17] is capable of generating TLINKs from SLINKs and Blinker – based on GutenLink [18] – contains a large set of relation postulations given configurations of EVENTs and TIMEXs and focuses on TLINKs.

Instead of prior versions of the toolkit which permitted co-operation of link annotation components via a voting mechanism [18]. TTK has a separate Link Merger component. The merger uses confidence scores from individual components as well as a pre-set bias (for example, to give low priority to the large number of classifier-generated links) to order candidate links. These are then sequentially tested against a temporal graph of the discourse, with consistency checking between each addition; inconsistent links are not added. This makes it impossible to revoke possibly incorrect information once it has been added, but generates a consistent annotation where high confidence is at least partially rewarded.

A.3.2 Callisto/Tango

TANGO is an assistive annotation tool that helps users build correct annotations from suggestions made by the included automatic temporal annotation systems, as well as a visual representation component. It is integrated within Callisto (Fig. A.2),

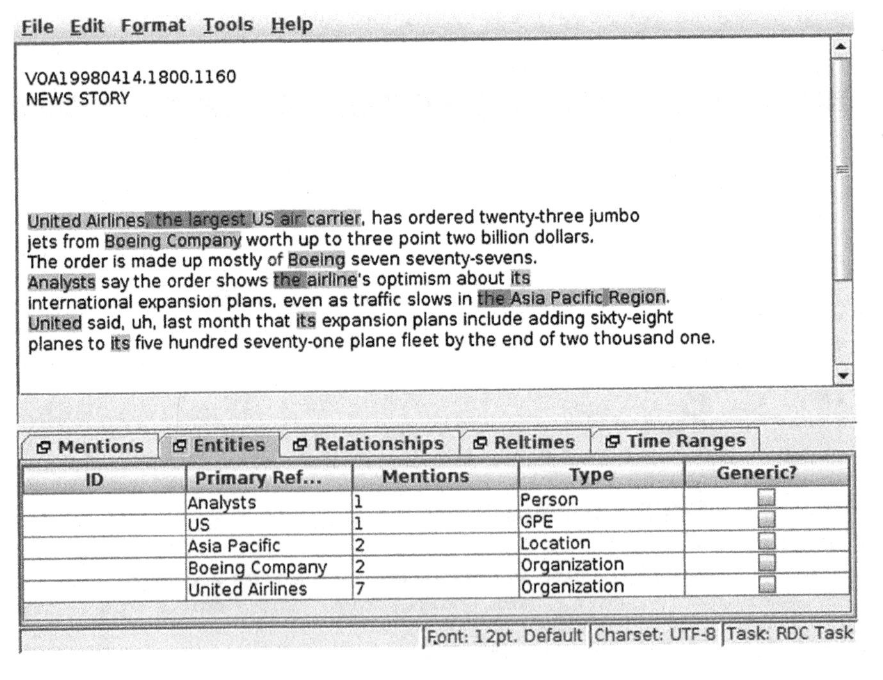

Fig. A.2 Manually annotating text with Callisto

a general-purpose manual linguistic annotation tool. Callisto's TANGO component for TimeML annotation is particularly strong for ease of temporal link annotation.

A.3.3 BAT

The Brandeis Annotation Tool, or BAT [19], enables collaborative semantic annotation and breaks down annotation into subtasks. It is a web-based tool, with administrator overview (see Fig. A.3). Multiple asynchronous and concurrent annotations can be made, making BAT a flexible tool for co-ordinating gold standard TimeML annotations. It has been used to create the TempEval-2 and Ita-TimeBank datasets.

A.3.4 Other Tools

Other purpose-built tools exist, such as Dante [20] which concentrates on temporal expression tagging and normalisation across many genres of text but is not are pub-

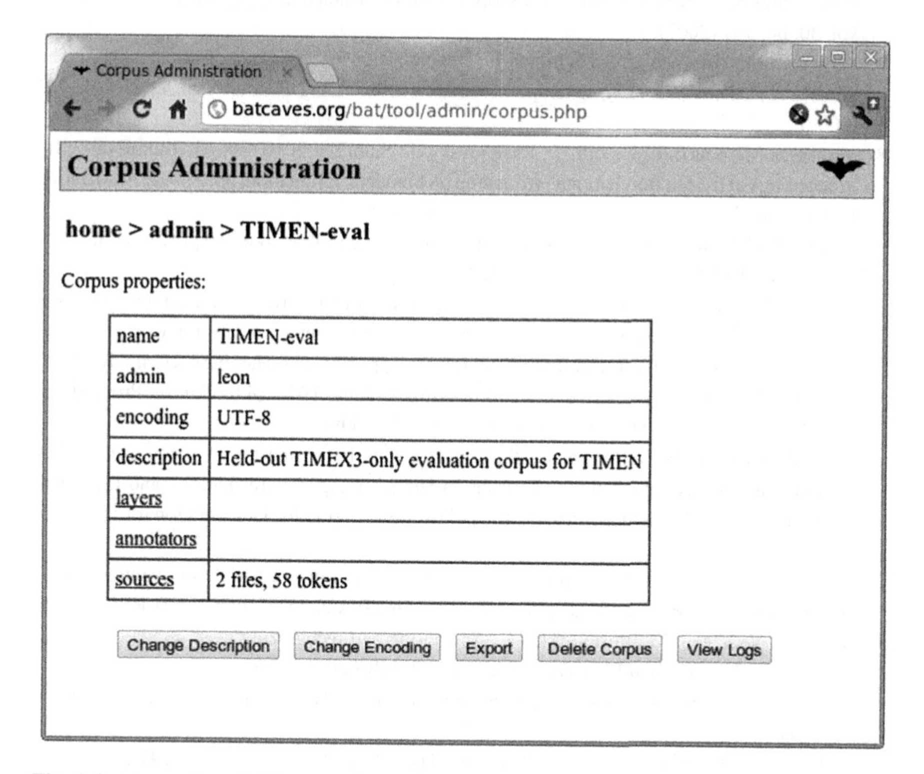

Fig. A.3 Overseeing a BAT annotation project

licly available. Existing general purpose language toolkits may also be adapted to cater for TimeML processing, such as NLTK [21], GATE [22] and Xara [23].

References

1. Pustejovsky, J., Hanks, P., Sauri, R., See, A., Gaizauskas, R., Setzer, A., Radev, D., Sundheim, B., Day, D., Ferro, L., et al.: The TimeBank Corpus. In: Proceedings of the Corpus Linguistics conference, pp. 647–656 (2003)
2. Boguraev, B., Pustejovsky, J., Ando, R., Verhagen, M.: TimeBank evolution as a community resource for TimeML parsing. Lang. Resour. Eval. **41**(1), 91–115 (2007)
3. Bergmeir, C., Benítez, J.: On the use of cross-validation for time series predictor evaluation. Inf. Sci. **191**, 192–2132 (2012)
4. Bittar, A., Amsili, P., Denis, P., Danlos, L.: French TimeBank: an ISO-TimeML annotated reference corpus. In: Proceedings of the 49th Annual Meeting of the Association for Computational Linguistics: Human Language Technologies: Short Papers, vol. 2, pp. 130–134. Association for Computational Linguistics (2011)
5. Caselli, T., Lenzi, V., Sprugnoli, R., Pianta, E., Prodanof, I.: Annotating events, temporal expressions and relations in Italian: the It-TimeML experience for the Ita-TimeBank. In: Proceedings of ACL-HLT 2011, p. 143. Association for Computational Linguistics (2011)
6. Forăscu, C., Ion, R., Tufiş, D.: Semi-automatic Annotation of the Romanian TimeBank 1.2. In: Proceedings of the RANLP 2007 Workshop on Computer-aided language processing (CALP), vol. 30, pp. 1–7 (2007)
7. Verhagen, M., Gaizauskas, R., Schilder, F., Hepple, M., Moszkowicz, J., Pustejovsky, J.: The TempEval challenge: identifying temporal relations in text. Lang. Resour. Eval. **43**(2), 161–179 (2009)
8. Verhagen, M., Saurí, R., Caselli, T., Pustejovsky, J.: SemEval-2010 task 13: TempEval-2. In: Proceedings of the 5th International Workshop on Semantic Evaluation, pp. 57–62. Association for Computational Linguistics (2010)
9. Mazur, P., Dale, R.: WikiWars: A new corpus for research on temporal expressions. In: Proceedings of the EMNLP, pp. 913–922 (2010)
10. Strötgen, J., Gertz, M.: WikiWarsDE: A German corpus of narratives annotated with temporal expressions. In: Proceedings of the Conference of the German Society for Computational Linguistics and Language Technology (GSCL 2011), pp. 129–134. Hamburg, Germany (2011)
11. Verhagen, M., Pustejovsky, J.: Temporal processing with the TARSQI toolkit. In: Proceedings of CoLing: Posters and Demonstrations, pp. 189–192 (2008)
12. Saurí, R., Knippen, R., Verhagen, M., Pustejovsky, J.: Evita: a robust event recognizer for QA systems. In: Proceedings of the conference on Human Language Technology and Empirical Methods in Natural Language Processing, p. 707. Association for Computational Linguistics (2005)
13. Saurí, R., Verhagen, M., Pustejovsky, J.: Slinket: A partial modal parser for events. In: Proceedings of the conference on Language Resources and Evaluation Conference (LREC) (2006)
14. Saurı, R., Verhagen, M., Pustejovsky, J.: Annotating and recognizing event modality in text. In: The 19th International FLAIRS Conference, FLAIRS 2006 (2006)
15. Verhagen, M.: Temporal closure in an annotation environment. Lang. Resour. Eval. **39**(2), 211–241 (2005)
16. Verhagen, M.: Drawing TimeML relations with TBox. Lect. Notes Comput. Sci. **4795**, 7 (2007)
17. Verhagen, M.: Times Between the Lines. Ph.D. thesis, Brandeis University (2004)

18. Verhagen, M., Mani, I., Sauri, R., Knippen, R., Jang, S., Littman, J., Rumshisky, A., Phillips, J., Pustejovsky, J.: Automating temporal annotation with TARSQI. In: Proceedings of the ACL 2005 on Interactive Poster and Demonstration Sessions, p. 84. Association for Computational Linguistics (2005)
19. Verhagen, M.: The Brandeis annotation tool. Lang. Resour. Eval. Conf. LREC **2010**, 3638–3643 (2010)
20. Mazur, P., Dale, R.: The DANTE temporal expression tagger. In: Human Language Technology. Challenges of the Information Society, p. 257. Springer, New York (2009)
21. Loper, E., Bird, S.: NLTK: The natural language toolkit. In: Proceedings of the ACL-02 Workshop on Effective Tools and Methodologies for Teaching Natural Language Processing and Computational Linguistics, vol. 1, pp. 63–70. Association for Computational Linguistics (2002). http://www.nltk.org/
22. Cunningham, H., Maynard, D., Bontcheva, K., Tablan, V., Aswani, N., Roberts, I., Gorrell, G., Funk, A., Roberts, A., Damljanovic, D., Heitz, T., Greenwood, M.A., Saggion, H., Petrak, J., Li, Y., Peters, W., Derczynski, L., et al.: Developing Language Processing Components with GATE Version 8 (a User Guide). University of Sheffield (2001). https://gate.ac.uk/releases/gate-8.2-build5482-ALL/doc/tao/tao.pdf
23. Burnard, L., Todd, T.: Xara: an XML aware tool for corpus searching. Proc. Corpus Linguist. **2003**, 142–144 (2003)

Appendix B
RTMML Reference

This appendix details extensions made to TimeML, so that it may capture extra information helpful for temporal reasoning, based upon Reichenbach's framework of tense and aspect [1].

B.1 Examples

B.1.1 Fiction

From *David Copperfield* by Charles Dickens:

Example 34 When he had put up his things for the night he took out his flute, and blew at it, until I almost thought he would gradually blow his whole being into the large hole at the top, and ooze away at the keys.

We give RTMML for the first five verbal events from Example 34 RTMML in Fig. B.1. The fifth, `v5`, exists in a context that is instantiated by `v4`; its reference time is defined as such. We can use one `link` element to show that `v2`, `v3` and `v4` all use the same reference time as `v1`. The temporal relation between event times of `v1` and `v2` can be inferred from their shared reference time and their tenses; that is, given that `v1` is anterior past and `v2` simple past, we know $E_{v1} < R_{v1}$ and $E_{v2} = R_{v2}$. As our `<rtmlink>` states $R_{v1} = R_{v2}$, then $E_{v1} < E_{v2}$. Finally, `v5` and `v6` happen in the same context, described with a second SAME_TIMEFRAME link.

B.1.2 Editorial News

From an editorial piece in TimeBank [2] (AP900815-0044.tml):

L.R.A. Derczynski, *Automatically Ordering Events and Times in Text*,
Studies in Computational Intelligence 677, DOI 10.1007/978-3-319-47241-6

```
<doc time="1850" mod="BEFORE" />
<!-- had put -->
<verb xml:id="v1"
  target="#range(#token2,#token3)"
  view="anterior" tense="past" />
<!-- took -->
<verb xml:id="v2" target="#token11"
  view="simple" tense="past" />
<!-- blew -->
<verb xml:id="v3" target="#token17"
  view="simple" tense="past" />
<!-- thought -->
<verb xml:id="v4" target="#token24"
  view="simple" tense="past" />
<!-- would gradually blow -->
<verb xml:id="v5"
  target="#range(#token26,#token28)"
  view="posterior" tense="past"
  se="=" er=">" sr=">"
  r="#v4.e" />
<!-- ooze -->
<verb xml:id="v6"
  target="#range(#token26,#token28)"
  view="posterior" tense="past"
  se="=" er=">" sr=">" />
<rtmlink xml:id="l1"
  type="SAME_TIMEFRAME">
  <link target="#v1" />
  <link target="#v2" />
  <link target="#v3" />
  <link target="#v4" />
</rtmlink>
<rtmlink xml:id="l2"
  type="SAME_TIMEFRAME">
  <link target="#v5" />
  <link target="#v6" />
</rtmlink>
```

Fig. B.1 RTMML for a passage from David Copperfield

Example 35 Saddam appeared to accept a border demarcation treaty he had rejected in peace talks following the August 1988 cease-fire of the eight-year war with Iran.

```
<doc time="1990-08-15T00:44" />
<!-- appeared -->
<verb xml:id="v1" target="#token1"
  view="simple" tense="past" />
<!-- had rejected -->
<verb xml:id="v2"
  target="#range(#token9,#token10)"
  view="anterior" tense="past" />
<rtmlink xml:id="l1"
  type="SAME_TIMEFRAME">
  <link target="#v1" />
  <link target="#v2" />
</rtmlink>
```

Here, we relate the simple past verb *appeared* with the anterior past (past perfect) verb *had rejected*, permitting the inference that the first verb occurs temporally after the second. The corresponding TimeML (edited for conciseness) is:

Example 36

```
Saddam <EVENT eid="e74" class="I_STATE">
appeared</EVENT> to accept a border demarcation treaty he had <EVENT eid="e77"
class="OCCURRENCE">rejected</EVENT>

<MAKEINSTANCE eventID="e74" eiid="ei1568"
 tense="PAST" aspect="NONE" polarity="POS"
 pos="VERB"/>
<MAKEINSTANCE eventID="e77" eiid="ei1571"
 tense="PAST" aspect="PERFECTIVE"
 polarity="POS" pos="VERB"/>
```

In this example, we can see that the TimeML annotation includes the same information, but a significant amount of other annotation detail is present, cluttering the information we are trying to see. Further, these two <EVENT> elements are not directly linked, requiring transitive closure of the network described in a later set of <TLINK> elements, which are omitted here for brevity.

B.1.3 Linking Events to Calendar References

RTMML makes it possible to precisely describe the nature of links between verbal events and times, via positional use of the reference point. We will link an event to a temporal expression, and suggest a calendrical reference for that expression, allowing the events to be placed on a calendar. Consider the below text, from wsj_0533.tml in TimeBank.

Example 37 At the close of business Thursday, 5,745,188 shares of Connaught and C$44.3 million face amount of debentures, convertible into 1,826,596 common shares, had been tendered to its offer.

```
<doc time="1989-10-30" />
<!-- close of business Thursday -->
<timerefx xml:id="t1"
  target="#range(#token2,#token5)" />
<!-- had been tendered -->
<verb xml:id="v1"
  target="#range(#token25,#token27)"
  view="anterior" tense="past" />
<rtmlink xml:id="l1" target="#t1 #v1">
  <link target="#t1" />
  <link target="#v1" />
</rtmlink>
```

This shows that the reference time of `v1` is `t1`. As `v1` is anterior, we know that the event mentioned occurred before *close of business Thursday*. Normalisation is not a task that RTMML addresses, but there are existing methods for deciding which Thursday is being referenced given the document creation date [3]; a time of day for *close of business* may be found in a gazetteer.

B.2 Annotation Notes

As can be seen in Table 6.2, there is not a one-to-one mapping from English tenses to the nine specified by Reichenbach. In some annotation cases, it is possible to see how to resolve such ambiguities. Even if view and tense are not clearly determinable, it is possible to define relations between S, E and R; for example, for arrangements corresponding to the simple future, $S < E$. In cases where ambiguities cannot be resolved, one may annotate a disjunction of relation types; in this example, we might say "$S < R$ or $S = R$" with `sr="<="`.

Contexts seem to have a shared speech time, and the $S - R$ relationship seems to be the same throughout a context. Sentences which contravene this (e.g. *"By the time I ran, John will have arrived"*) are rather awkward. Contexts are typically broken by timexes (e.g. positional use of the reference point), shifting of frame of reference by use of *"then"*, use of temporal signals or any boundary of reported speech (e.g. starting and ending quotes).

RTMML annotation is not bound to a particular language. As long as a segmentation scheme (e.g. WordSeg-1) is agreed and there is a compatible system of tense and aspect, the model can be applied and an annotation created.

References

1. Reichenbach, H.: The tenses of verbs. In: Elements of Symbolic Logic. Dover Publications, New York (1947)
2. Pustejovsky, J., Hanks, P., Sauri, R., See, A., Gaizauskas, R., Setzer, A., Radev, D., Sundheim, B., Day, D., Ferro, L., et al.: The TimeBank Corpus. In: Proceedings of the Corpus Linguistics conference, pp. 647–656 (2003)
3. Mazur, P., Dale, R.: Whats the date? High accuracy interpretation of weekday. In: 22nd International Conference on Computational Linguistics (Coling 2008), Manchester, UK, pp. 553–560 (2008)
4. Derczynski, L., Gaizauskas, R.: An Annotation Scheme for Reichenbach's Verbal Tense Structure. In: Proceedings of the 6th Joint ACL-ISO Workshop on Interoperable Semantic Annotation (ISA 6), Oxford, UK (2011)
5. Derczynski, L., Gaizauskas, R.: Empirical Validation of Reichenbach's Tense Framework. In Proceedings of the 10th Conference on Computational Semantics (IWCS), Potsdam, Germany (2013)

Appendix C
CAVaT Reference

This section contains a reference guide for the CAVaT package [1]. Up to date information can always be found at https://github.com/leondz/cavat.

C.1 Installation and Configuration

The first time CAVaT is run, it will attempt to create a directory `$HOME/.cavat/`, where it will store its SQLite files.

C.2 Getting Started

Enter the following to load a TimeML corpus into the "test" database - it's important to include the trailing slash / in the path:

```
cavat> corpus import /home/user/corpus/data/ to test
```

Depending on your disk and CPU speeds, this might take about 10–20 seconds per megabyte of TimeML. If it seems to take longer, you can get more information about what CAVaT is doing during import by enabling debug mode before import:

```
cavat> debug on
```

Leave debug mode with a simple:

```
cavat> debug off
```

Once the corpus has loaded, you can use `corpus info` to see metadata about the import, or `corpus list` to see an available list of corpora. To switch between corpora, and to select a newly loaded one, enter `corpus use <name>`.

L.R.A. Derczynski, *Automatically Ordering Events and Times in Text*,
Studies in Computational Intelligence 677, DOI 10.1007/978-3-319-47241-6

C.3 Queries

The show command is used for generating reports on the currently loaded corpus. Reports focus on one tag type, and give information about their attributes. One can view all values for a tag with `list` reports, or the distribution of values with `distribution` reports, or simply see how many instances of that tag list a value for a field with `state` reports.

Reports can be provided in multiple formats; there is:

- `screen` - for screen or fixed-width font output
- `csv` - comma separated values
- `tex` - LaTeX table format

The general format for report generation is:

```
show <report type> of <tag>  <field> [as <format>]
```

To try a simple query, enter:

```
cavat> show distribution of tlink reltype
```

You should see a table listing the values listed for relType in the current corpus' TLINK tags, as well as their frequencies. To see how many TLINKs use a signal, and use the result in a LaTeX document, you can try:

```
cavat> show state of tlink signalid as tex
```

Reference

1. Derczynski, L., Gaizauskas, R.: Analysing temporally annotated corpora with CAVaT. In: Proceedings of the Language Resources and Evaluation Conference, pp. 398–404 (2010)

Index

L.R.A. Derczynski, *Automatically Ordering Events and Times in Text*,
Studies in Computational Intelligence 677, DOI 10.1007/978-3-319-47241-6

Printed in the United States
By Bookmasters